Bioreactor and Ex Situ Biological Treatment Technologies

Editors

Bruce C. Alleman
and Andrea Leeson
Battelle

The Fifth International In Situ and
On-Site Bioremediation Symposium

San Diego, California, April 19–22, 1999

BATTELLE PRESS
Columbus • Richland

Library of Congress Cataloging-in-Publication Data

Bioreactor and ex situ biological treatment technologies / editors, Bruce C. Alleman and
    Andrea Leeson
        p.   cm.
        Proceedings from the Fifth International In Situ and On-Site Bioremediation
    Symposium, held April 19–22, 1999, in San Diego, California.
        Includes bibliographical references and index.
        ISBN 1-57477-078-0 (hardcover : alk. paper)
        1. Bioremediation Congresses. 2. Bioreactors Congresses.
    I. Alleman, Bruce C., 1957– . II. Leeson, Andrea, 1962– .
    III. International Symposium on In Situ and On-Site Bioremediation
    (5th : 1999 : San Diego, Calif.)
    TD192.5.B556   1999
    628.5--dc21   99-23338
                                                                                  CIP

Printed in the United States of America

Battelle Press
505 King Avenue
Columbus, Ohio 43201, USA
614-424-6393 or 1-800-451-3543
Fax: 1-614-424-3819
Internet: press@battelle.org
Website: www.battelle.org/bookstore

For information on future environmental conferences, write to:
    Battelle
    Environmental Restoration Department, Room 10-123B
    505 King Avenue
    Columbus, Ohio 43201-2693
    Phone: 614-424-7604
    Fax: 614-424-3667
    Website: www.battelle.org/conferences

CONTENTS

FOREWORD

The Fifth International In Situ and On-Site Bioremediation Symposium was held in San Diego, California, April 19–22, 1999. The program included approximately 600 platform and poster presentations, encompassing laboratory, bench-scale, and full-scale field studies being conducted worldwide on a variety of bioremediation and supporting technologies used for a wide range of contaminants.

The author of each presentation accepted for the program was invited to prepare a six-page paper, formatted according to specifications provided by the Symposium Organizing Committee. Approximately 400 such technical notes were received. The editors conducted a review of all papers. Ultimately, 389 papers were accepted for publication and assembled into the following eight volumes:

Natural Attenuation of Chlorinated Solvents, Petroleum Hydrocarbons, and Other Organic Compounds – Volume 5(1)
Engineered Approaches for In Situ Bioremediation of Chlorinated Solvent Contamination – Volume 5(2)
In Situ Bioremediation of Petroleum Hydrocarbon and Other Organic Compounds – Volume 5(3)
Bioremediation of Metals and Inorganic Compounds – Volume 5(4)
Bioreactor and Ex Situ Biological Treatment Technologies – Volume 5(5)
Phytoremediation and Innovative Strategies for Specialized Remedial Applications – Volume 5(6)
Bioremediation of Nitroaromatic and Haloaromatic Compounds – Volume 5(7)
Bioremediation Technologies for Polycyclic Aromatic Hydrocarbon Compounds – Volume 5(8)

Each volume contains comprehensive keyword and author indices to the entire set.

This volume focuses on the use of engineered systems to treat vapor, water, slurries, and soils. Many site- and wastestream-remediation challenges are best met by applying bioreactors and other ex situ treatment technologies, sometimes singly and sometimes in combination with other technologies. The articles in this volume cover a number of treatment technologies—slurry reactors, compost reactors, biopiles, packed-bed reactors, vapor-phase reactors, and biofilters—used to remediate a variety of media affected by contaminants such as arsenic, TPHs, MTBE, heavy metals, solvents, and other wastes.

We would like to thank the Battelle staff who assembled the eight volumes and prepared them for printing. Carol Young, Lori Helsel, Loretta Bahn, Gina Melaragno, Timothy Lundgren, Tom Wilk, and Lynn Copley-Graves spent many hours on production tasks—developing the detailed format specifications sent to each author; examining each technical note to ensure that it met basic page layout requirements and making adjustments when necessary; assembling the volumes; applying headers and page numbers; compiling the tables of contents and author and keyword indices, and performing a final check of the pages before submitting them to the publisher. Joseph Sheldrick, manager of Battelle Press, provided valuable production-planning advice and coordinated with the printer; he and Gar Dingess designed the covers.

The Bioremediation Symposium is sponsored and organized by Battelle Memorial Institute, with the assistance of a number of environmental remediation organizations. In 1999, the following co-sponsors made financial contributions toward the Symposium:

Celtic Technologies   U.S. Microbics, Inc.
Gas Research Institute (GRI)   U.S. Naval Facilities Engineering
IT Group, Inc.   Command
Parsons Engineering Science, Inc.   Waste Management, Inc.

Additional participating organizations assisted with distribution of information about the Symposium:

Ajou University, College of   U.S. Air Force Center for
   Engineering   Environmental Excellence
American Petroleum Institute   U.S. Air Force Research Laboratory
Asian Institute of Technology   Air Base and Environmental
Conor Pacific Environmental   Technology Division
   Technologies, Inc.   U.S. Environmental Protection
Mitsubishi Corporation   Agency
National Center for Integrated   Western Region Hazardous
   Bioremediation Research &   Substance Research Center
   Development (University of   (Stanford University and
   Michigan)   Oregon State University)

The materials in these volumes represent the authors' results and interpretations. The support of the Symposium provided by Battelle, the co-sponsors, and the participating organizations should not be construed as their endorsement of the content of these volumes.

Bruce Alleman and Andrea Leeson, Battelle
1999 Bioremediation Symposium Co-Chairs

SOIL AND WATER REMEDIATION AROUND THE WAREHOUSE IN BORNE SULINOWO

Jerzy Bil, Andrzej Spychała Military University of Technology, Warsaw, Poland
Andrzej Rybka EXBUD, Kielce, Poland, *Joanna Surmacz-Górska* Silesian
Technical University, Gliwice, Poland

ABSTRACT: In this paper, an in situ method for soil and water environment purification within a chemical storage area is presented. The chemical storage area is located in Borne Sulinowo and was used by Soviet Union military troops between 1945 to1992. This region had been contaminated with polycyclic aromatic hydrocarbons (PAHs), organic chlorine compounds, aliphatic derivatives of phthalates, etc. In this area, aliphatic derivatives of phosphoric acid were found which indicates the possible contamination of this area with toxic warfare agents. Results after 6 months of remediation have been presented.

INTRODUCTION

The site near Borne Sulinowo, which is approximately 18,000 ha in area, is a post-German military training ground for artillery, which had been taken over in 1945 and used by the Soviet Union troops up to 1992. On the northern side, the site is adjacent to Pile Lake, whose surface area is approximately 1,000 ha. On the south bank of the lake is the town of Borne Sulinowo, which covers an area of approximately 1,800 ha and has 2,500 inhabitants.

The complex of warehouses included a separate and specially protected, expanded area with a surface of 6 ha, where chemicals connected with utilization of military techniques and possibly also toxic warfare agents were stored and distributed. The research conducted on the grounds of the chemical warehouse in 1993 to 1995 located 6 areas having a total surface of 1.3 ha, on which excessive pollution was discovered. The area on which the distribution of various chemical compounds took place was especially dangerous. This paper will present a method and the results of cleaning the environment in this particular area.

Geological Structure and Threads. The area under consideration is composed of Tertiary and Quaternary formations. Tertiary formations are composed of medium-grain quartz sands found at the level of 49.0 to 55.0 m below ground level as well as mudstone and green-gray mudstone drilled at a depth of 55.0 to 60.0 m and not rebored. The Quaternary formations are composed of sands of various grain sizes with gravel and mixed interbeddings within the chemical storage area. Groundwater is present at 8 m below ground level. On the basis of measurements made in 1993 to 1995, we may conclude that the fluctuation amplitude of the groundwater level is approximately 1 m. A groundwater velocity of about 80 m per year was measured in the direction of Pile Lake, which is located 1.2 km from the area. Thus the chemical storage area is situated within the area flow of this lake. In accordance with site planning of the town and commune of Borne Sulinowo, the storage area is intended for economic activity.

In accordance with the regulations (Methodical guidelines, 1994), admissible concentrations of impurities in soil and in groundwater may not exceed the values shown in Table 1 determined for this area. Additionally, State Environment Protection Inspection assumed that in the depth zone of 0 to 2 m below ground level, the impurities concentration value may not exceed one half of these values (0.5 C) (Table 1). Within the area under discussion, concentrations significantly exceeding the admissible values were found. Most frequently the military operation materials, petroleum-based products as well as disinfectants, were the sources of these impurities. Additionally, trace quantities of aliphatic derivatives of phosphoric acid were found in single water and soil specimens, which show that the terrain has probably been contaminated with phosphor-organic toxic warfare agents.

The characteristic feature of the remediation region discussed here was large quantities of various impurities located at different depths. Therefore, it was assumed the preliminary examinations that had been performed did not determine all sources present within this area. Thus, we expected that the remediation process would uncover unrecognized impurities, which was the case.

The overall goal of the remediation process was to remove impurities and to return to the town the attractive terrain of 6 ha in size.

MATERIALS AND METHODS

Land Remediation System. The in situ system of "pump and treat" was used to clean the area. The basic elements of the system were:
  a) water installation
  b) air installation
  c) container water treatment plant

Water Installation. Water installation is composed of the following elements:
  a) 2 engineering wells 160 mm in diameter and 15 m deep. The purpose of the well was to extract contaminated water and to supply it to the water treatment plant. By pumping the water, the proper depression is produced, which prevents infiltration of water impurities beyond the purification region;
  b) 2 pressing pipelines provided water from the well to the water treatment plant, of a total length of 110 m, connected into one pressing pipeline before the plant;
  c) 50 m long horizontal drainage used to strain purified water in the ground, located from the side of groundwater run-off to increase hydraulic gradient;
  d) A set for spraying the reclamation surface with purified water.

Air Installation. The air installation is composed of the following elements:

TABLE 1. Organic compounds content in samples of water and soil taken on area of former chemical materials store in Borne Sulinowo in purification process.

Ord. No.	Type of contamination	Contamination condition of underground water [µg/dm³]						Contamination condition of soil [mg/kg d.m.]					
		Admissible "C"	Origin condition	After 6 months	After 12 months	After 15 months	After 18 months	Admissible 1/2"C"	Origin condition	After 6 months	After 12 months	After 15 months	After 18 months
1.	Aromatic hydrocarbons as well as their derivatives - sum	100	948	144	134	98	29	135	0	705	580	140	125
2.	Polycyclic aromatic hydrocarbons - sum	40	0	44	40	39	38	100	559	376	282	110	100
3.	Chlorinated hydrocarbons, organic chlorine compounds - sum	5	0	0	0	0	0	10	0	440	167	0	0
4.	The remaining contamination's (aliphatic, cyclic hydrocarbons as well as their derivatives, aliphatic derivatives of phthalates, hydroxyl derivatives of upper hydrocarbons	150	482	458	320	130	95	400	471	355	378	155	95
5.	Aliphatic derivatives of phosphoric acid		1	1.6		0		0.5	0.5	0.2	0.1	0	

a) 2 engineering piezometers, 11 m deep, used to aerate the ground. In cases where impurities occur in them, it is possible to pump water to the treatment plant,

b) 2 vertical drainpipes, 10 m deep, to suck off the soil gas. The soil gas, after purification through carbon, is emitted to the atmosphere;

c) 200 m long horizontal drainage system provided with alternate aeration of soil as well as extracting soil gas, which after purification through carbon, is also emitted to the atmosphere.

Container Water Treatment Plant. The container water treatment plant is fully equipped to remove chemical compounds, petroleum-based products, as well as purifying extracted soil gas (Figure 1). The system is adapted for continuos operation in winter and in summer within the temperature range of -20°C up to +40°C. Contaminated water is pumped from the well and supplied to the separator. In this container, the water is averaged, separation of suspended matters is carried out, and possible petroleum-based products included in drawn water. The separator operates on the mechanical basis utilizing the hydrodynamic coalescence effect. Then water is pumped to the stripping chamber in which iron and manganese compounds are removed by means of aeration. This chamber has a warm air blow from the bottom and in the upper part a fan, which extracts the hydrocarbon vapors through a carbon filter. From equalizing tank located under the stripping chamber, water is pumped onto the filter, whose task is to remove iron and manganese from water as well as optimizing the pH factor. It is equipped with an aeration system. This is the second stage of iron and manganese removal. It provides high aeration of water before it is supplied to the decarbonate bed as well as trivalent iron is caught on the bed. Washings from deironing equipment are cleaned in a special device beyond the container. Water flows under pressure to columns filled with mineral sorbents and activated carbon, where hydrocarbons are removed. The flowrate of water through the columns is adapted to optimum time of bed effect. At the effluent of each sorbent column there is a control system controlling the amount of effluent to the container of cleaned water, which is constructed in the container at the end of columns set.

Agrotechnical Procedure. To support the bioremediation process, the following activities were carried out periodically (twice a year):
a) deep ploughing and harrowing,
b) sowing fertilizer,
c) sowing of lupine.

RESULTS AND DISCUSSION
Within the storage area, an area of 1.3 ha and 60,000 m^3 in volume had been contaminated. Groundwater had been contaminated as well. The contaminated area and volume are 0.3 ha and 11 thousand m^3, respectively. Detailed information on impurities and their concentrations are presented in Table 1.

The remediation installation took about two months. During this time period, the final verification of the design assumption of the plant was carried out as well as the optimum output amounting to approximately 70 m^3 per 24 hours. This value ensured the proper utilization of the filtration bed of the plant and the whole area of depression was also covered. Within the period of start up, approximately 2,100 m^3 of water was purified. The complete operation of the system was accomplished in 1996/1997. About 42,000 m^3 of water was purified by the end of June 1998. Filtering of purified water was carried out in winter by means of horizontal draining pipe system mounted below the frost penetration level of the ground, whereas in summer it was carried out by spraying the surface.

The effects of environment purification after 18 months of continuos operation of the system are presented in Table 1. Analysis of these results shows that the first six months of system operation led to deterioration of the environment condition. This effect was expected. It was assumed that within the first period of purification process, unidentified sources of contamination would be discovered. Therefore, the first six months of system operation could determine relatively well the initial condition of contamination (initial conditions of the process). Further results of the examination show that the purification system had entirely fulfilled its task. After 18 months of operation the results which were obtained allowed us to complete the purification process.

RESULTS

1. The region under consideration is an atypical example of diversely contaminated environment, with local sources, located at different depths;
2. Complete characterization of an environment in such a situation is in practice impossible. It would be necessary to take several thousand samples for examination. The reclamation process was based on initial conditions which were determined inaccurately;
3. The first time period of continuos operation of the system lasted about 6 months and revealed unidentified contamination sources and determined relatively precisely the initial conditions of the purification process;
4. The container plant designed for needs of purification of the chemical storage area had fully fulfilled its task because it removed about 90% of the contamination. The depression produced by pumps located in engineering wells covered the whole area, which guaranteed non-spreading of contamination beyond the boundaries of this area;
5. Complete purification of the whole region was made after 18 months of system operation. The intended effects were achieved and the system was switched off from operation on 30th June 1998;
6. Reclamation cost for 1 m^3 of soil amounted to about 6 US dollars.

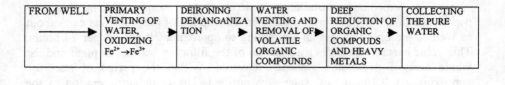

FROM WELL	PRIMARY VENTING OF WATER, OXIDIZING $Fe^{2+} \rightarrow Fe^{3+}$	DEIRONING DEMANGANIZATION	WATER VENTING AND REMOVAL OF VOLATILE ORGANIC COMPOUNDS	DEEP REDUCTION OF ORGANIC COMPOUDS AND HEAVY METALS	COLLECTING THE PURE WATER

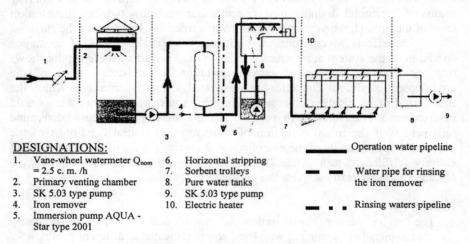

DESIGNATIONS:

1. Vane-wheel watermeter Q_{nom} = 2.5 c. m. /h
2. Primary venting chamber
3. SK 5.03 type pump
4. Iron remover
5. Immersion pump AQUA - Star type 2001

6. Horizontal stripping
7. Sorbent trolleys
8. Pure water tanks
9. SK 5.03 type pump
10. Electric heater

————————— Operation water pipeline

— — — — Water pipe for rinsing the iron remover

—— • —— •• Rinsing waters pipeline

FIGURE 1. Water treatment plant operation system

REFERENCE

Methodical guidelines related to soil and sub-soil water contamination rate by means of petroleum-derived products and other chemical substances within reclamation process. (Polish) State Environment Protection Inspection, Warsaw, 1994.

A COMBINED REMOVAL OF OIL, TOXIC METALS AND ARSENIC FROM CONTAMINATED SOILS

Veneta I. Groudeva (University of Sofia, Sofia, Bulgaria)
Stoyan N. Groudev (University of Mining and Geology, Sofia, Bulgaria)
Ilyana A. Ivanova (University of Sofia, Sofia, Bulgaria)

ABSTRACT: In the Tulenovo oil deposit, Northeastern Bulgaria, soils have been heavily contaminated with oil, toxic heavy metals (mainly lead and cadmium) and arsenic. A pilot scale operation for a combined removal of these pollutants from the soils was carried out using the heap technique. The soil remediation was based on the activity of both indigenous and laboratory-bred microorganisms which formed a stable community.

The oil was degraded to non-toxic products, while the heavy metals and arsenic were solubilized mainly as complexes with secreted microbial metabolites and were removed from the heap by drainage waters percolating through the soil mass. The regular ploughing up of the soil, the injection of air, the addition of nutrients (ammonium and phosphate), the maintenance of water content in the range of about 30-35 % and temperature of the soil were the most essential factors for enhancing the microbial growth and activity. The oil content in the heap was decreased from the initial 20-25 to less than 2 g/kg dry soil within 8 months of treatment (from the middle of March 1997 to the middle of November 1997). Simultaneously, the contents of lead, cadmium and arsenic were decreased below the relevant permissible levels.

INTRODUCTION

The soils in the Tulenovo oil deposit, Northeastern Bulgaria are polluted with crude oil. The oil is heavy, with a specific gravity of 0.939, rich in asphaltene-resinous substances and with a high viscosity. Since 1990 bioremediation of soils is applied using the heap technique (Groudeva et al., 1993; Groudeva et al., 1994). Some field sections in the deposit represent a specific case because apart from the oil, they are polluted with toxic heavy metals and arsenic. Recently, a pilot-scale operation for a combined removal of oil and toxic metals from the soils was started using an experimental heap consisting of a heavily polluted soil. Some data about this operation are shown in this paper.

EXPERIMENTAL

The heap had the shape of a truncated pyramid and was formed on a ground surface covered by a clay layer. The heap was about 20 m long, 0.7 m high, 5 m wide at the top end and 7 m wide at the bottom end, and contained about 130 tons of soil. The initial oil content in the heap was about 20-25 g/kg dry soil. The contents of some toxic heavy metals in the heap (Pb, Cd, Cu) as well as of arsenic were higher than the relevant permissible levels.

The polluted soil contained their own indigenous microflora including different oil-degrading and metal-solubilizing microorganisms. It was found, however, that the inoculation of the soil by a mixed laboratory-bred microbial culture consisting of several very active oil-degrading microorganisms (related to the genera Bacillus, Rhodococcus, Pseudomonas and Corynebacterium) enhanced considerably the rate of oil degradation and, at the same time, had no negative effect on the rate of metal solubilization. This laboratory-bred culture possessed the ability to degrade at relatively high rates the very high molecular weight asphaltenes, resins, and complex aromatics which, even under optimum conditions for microbial growth, were degraded very slowly by the indigenous microorganisms. After the inoculation, the introduced microorganisms rapidly formed a stable community with the indigenous microflora. Each member of this community possessed specific enzymes and was able to degrade only some of the many types of molecules in the complex hydrocarbon mixture. However, the microbial community as a whole demonstrated a well expressed synergetic action which realized an efficient degradation of the oil hydrocarbons in the soil.

The toxic heavy metals and arsenic were solubilized from some easily leachable fractions of the soil mainly by means of microbially secreted organic acids. It must be noted, however, than these metals were solubilized also, although at very low rates, even from the relevant sulphide minerals present in the soil. This was due to the activity of some chemolithotrophic bacteria, mainly such related to the species Thiobacillus thioparus and Thiobacillus neapolitanus. These bacteria grow in media with neutral, slightly alkaline or slightly acidic pH values and enhance the oxidation of sulphide minerals by removing the passivation films of $S°$ deposited on the mineral surface as a result of different chemical, electrochemical and biological processes (Groudev, 1990).

The treatment of the soil was connected with a regular ploughing down to the clay layer to enhance the oil deposit to maintain its water content in the range of 30-35 %. Zeolite saturated with ammonium phosphate was added to the soil (in amounts in the range of 2-5 kg/ton dry soil) to provide the microorganisms with a suitable source of nitrogen and phosphorus and to improve the physico-mechanical properties of the soil. The concentrations of total ammonia within soil pore waters was monitored and maintained below the established permissible level for water used in agriculture and/or industry.

Periodically, the soil was flushed with water to remove products from the soil degradation as well as the soluble toxic heavy metals and arsenic. A system to collect the soil effluents was constructed around the heap. These effluents were treated in a natural wetland located near the heap.

The rate of oil biodegradation markedly depended on the temperature of the soil. About 90% of the oil was degraded within 8 months (from the middle of March to the middle of November) at temperatures varying from $+3°C$ to $+37°C$ (Table 1). The toxic heavy metals and arsenic were also efficiently removed from

TABLE 1. Biodegradation of the different hydrocarbon components of the soil in the soil heap

Hydrocarbon component	Content in the oil, %		Content in the soil, g/kg dry soil		Hydrocarbon degradation, %
	I	II	I	II	
Parafins	20.3	0.5	4.9	0.01	99.8
Naphthenes	58.1	4.4	13.9	0.08	99.4
Aromatics + polars	21.6	95.1	5.2	1.71	67.1
Total	100.0	100.0	24.0	1.80	92.5

Notes: I - Before treatment; II - After treatment;
The hydrocarbons were determined by direct extraction from the relevant soil sample with 1,1,2 -trifluoroethane and IR-determination.

TABLE 2. Removal of toxic elements from the experimental soil heap

Toxic elements	Content in the soil, mg/kg dry soil		Pollutant removal, %	Permissible levels for soils with pH >7.0
	Before treatment	After treatment		
Pb	208	60	71.2	8 0
Cd	5.5	2.1	61.8	3.0
Cu	305	145	52.5	28 0
Zn	284	127	55.3	370
As	35	14	60.0	25

the soil and at the end of the treatment their contents were lower than the relevant permissible levels (Table 2).

The analysis of the microflora in the heap showed that the total number of aerobic heterotrophic bacteria as well as the number of oil-degrading microorganisms steadily increased during the treatment and in the period from May to October exceeded 10^9 and 10^8 cells/g dry soil, respectively (Table 3). Apart from the microorganisms, some higher organisms (protozoa, worms, insects, etc.) inhabited the soil and their numbers increased simultaneously with the reduction of the oil and toxic metals contents of the soil.

TABLE 3. Concentrations of various physiological groups of microorganisms in the experimental soil heap

Microorganisms	Cells /g dry soil	
	At the start of the treatment (March 1997)	In the middle of the treatment (July 1997)
Oil-degrading microorganisms	3.10^5	3.10^8
Aerobic heterotrophic bacteria	8.10^7	1.10^9
Oligocarbophiles	8.10^5	2.10^7
Cellulose-degrading microorganisms	1.10^5	3.10^6
Nitrogen-fixing bacteria	1.10^5	3.10^6
Nitrifying bacteria	4.10^4	3.10^6
Chemolithotrophic sulfur-oxidizind bacteria	7.10^4	8.10^5
Anaerobic heterothrophic bacteria	9.10^5	3.10^6
Denitrifying bacteria	8.10^4	8.10^5
Anaerobic bacteria, fermenting carbohydrates with gas production	8.10^4	9.10^5
Sulfate-reducing bacteria	6.10^4	5.10^5
Streptomycetes	3.10^5	1.10^6
Fungi	1.10^5	6.10^5
Total cells number	9.10^7	2.10^9

Note: The quantitative determination of the different physiological groups of the microorganisms was carried out by the spread plate technique on solid nutrient media or by the most probable number method. The total number of microbial cells was determined by epifluorescence microscopy.

CONCLUSION

The data from the above-mentioned experiment showed that it is possible to achieve an efficient simultaneous removal of oil, toxic metals and arsenic from heavily polluted soils using a bioremediation heap technique. The construction of a system to collect the soil effluents is necessary for the application of such treatment. A system to treat the soil effluents containing the removed pollutants is also needed. This can be carried out efficiently by means of a suitable passive system.

REFERENCES

Groudev, S. N. 1990. "Microbiological Transformations of Mineral Raw Materials". Doctor of Biological Sciences Thesis (in Bulgarian), University of Mining and Geology, Sofia.

Groudeva,V. I., S.N. Groudev, and I.A.Ivanova. 1993. "Microbial remediation of oil-polluted soils in the Tulenovo oil deposit, Bulgaria". In C.A.Jerez, T.Vargas, H.Toledo and J.V. Wiertz (Eds), Biohydrometallurgical Processing, vol.II, pp. 393-399.

Groudeva,V. I., S. N. Groudev, G. C. Uzunov and I. A. Ivanova. 1994. "Microbial removal of oil from polluted soils in a pilot scale operation in Tulenovo deposit, Bulgaria". In D.S. Holmes and R.W. Smith (Eds.), Minerals Bioprocessing II, pp. 231-240. TMS Minerals, Metals & Materials Society.

BIOREMEDIATION OF OIL-CONTAMINATED SOIL IN KUWAIT

Hiroyuki CHINO, Hirokazu TSUJI, Takashi MATSUBARA
(Obayashi Corporation, Tokyo, Japan)
Nader AL-AWADHI, M. Talaat BALBA, Reyad AL-DAHER
(Kuwait Institute for Scientific Research, Safat, Kuwait)

ABSTRACT: This paper summarizes the results of a 1-ha large-scale bioremediation field experiment of oil-contaminated soil in Kuwait's desert. Three techniques were used: landfarming, windrow composting soil piles and static piles fitted with an enforced aeration system. Total petroleum hydrocarbon (TPH) values decreased rapidly during the first 4 mo and then more gradually over the subsequent months. After 15 mo, 75 to 80% of the TPH in all the plots and piles had been degraded. More than 80% of 3-ring PAHs and 50% of 4 to 5-ring PAHs were degraded after 12 mo of treatment. Of the three methods tested, the landfarming method achieved in the highest decomposition rate and resulted in the greatest reduction in bioremediation time. Plant toxicity experiments were carried out using 15 mo bioremediated soil. The yield was almost within the normal range, so the treated soil had been sufficiently restored for vegetation. From the plant analysis after vegetation, it was clear that the uptake of PAH, sulfur and heavy metals were negligibly small. This study was carried out under the Joint Research Program between the Petroleum Energy Center under the Ministry of International Trading and Industry, Japan, and the Kuwait Institute for Scientific Research, Kuwait.

INTRODUCTION

During the Arabian Gulf War in 1991, Iraqi forces bombed and set fire to over 600 of Kuwait's oil wells, spilling enormous amounts of crude oil. The oil that flowed out above ground created over 300 oil lakes, covering a combined area of over 49 km^2 area. The remediation of the oil lake beds was actively considered so that oil contamination does not pose a critical health hazard to man and also to stimulate the restoration of the damaged ecosystem. Bioremediation technology was selected for the treatment of the contaminated soil, and three different methods were evaluated on the field.

MATERIALS AND METHODS

Site Preparation and Construction. A field demonstration experiment was carried out at the site near oil lake No. 102, located about 70 km south of the city of Kuwait. One hectare of this lake was designated for the project. The excavated soil was moved to the treatment yard, then homogenized well. TPH concentration in moderately contaminated soil was about 4%, while that in lightly contaminated soil was about 2%. TABLE 1 provides some information on the chemical and physical characteristics of the contaminated soil. The high salinity observed in the soil was caused by the large volume and subsequent evaporation of seawater used to extinguish the oil well fires. Nitrogen and phosphate fertilizers were added to the soil. Additionally, wood chips and compost were mixed into the soil at a rate of 5% of the soil volume. Three different types of bioremediation technologies were used: (a) landfarming for the treatment of 1,500 m³, of moderately and lightly contaminated soil; (b) windrow composting piles, fitted with an irrigation system, used for the treatment of 500 m³, of moderately contaminated soils; and (c) static soil piles, fitted with irrigation and enforced aeration systems, for the treatment of 250 m³ of lightly contaminated soils. FIGURE 1 illustrates the conceptual design and layout of these three treatment technologies. The soil's moisture content was maintained at between 8 and 10%. The bioremediation field demonstration started in July 1995, and continued until September 1996.

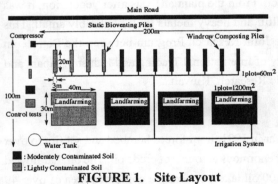

FIGURE 1. Site Layout

Monitoring. The items monitored in the tests are listed in TABLE 2. The soil's moisture content and temperature were measured every day; TPH, inorganic compounds, and microbial count were measured once every month. In addition, the quantities of oil component fractions in the total extractable matter (TEM), the aliphatic series fraction and the 16 types of hazardous aromatic compounds were analyzed in detail by GC and HPLC once every three months. Samples were collected from depths of approximately 10 and 30 cm from 30 evenly spaced spots in each of the land-

TABLE 1. Soil Characteristics

		Moderately Cont. Soil	Lightly Cont. Soil	Clean Desert Soil
Moisture Content	(%)	2.81	1.92	1.79
Conbustible Matter	(%)	9.03	5.06	2.25
TPH	(%)	3.94	1.87	0.02
TEM	(%)	6.15	2.4	0.03
pH		7.1	7.7	9.2
EC(Soil:Water=1:2	(mS/cm)	14.2	6.9	0.22
Chloride	(mg/kg)	6283	2530	153
Grain Size (40%)	(mm)	0.2	0.2	0.2
Grain Size (10%)	(mm)	0.1	0.1	0.1

TABLE 2. Items for Analysis and Methods

Item	Methods
Moisture	105℃, 1 night
Amount of Irrigation	Flowmeter
Ambient Temperature	Automated measurement by data logger
Soil Temperature	
TPH	Freon extraction, measured by FTIR (EPA 418.1)
TEM	Dichloromethane extraction, gravimetrically
Fractional Analysis	Column chromatography
Aliphatic Compounds	Freon extraction, GC (FID)
Aromatic Compounds	Dichloromethane extraction, HPLC (FLD)
pH	Soil :Water =1:2, by pH meter
Combustible Matter	550℃ 3 h
EC	Soil :Water =1:2, by EC meter
Chloride	AgCl, titration method
Sulphate	$BaSO_4$, nephlometric method
Phosphate	Molybdenum yellow colorimetric method
Ammonium Nitrogen	Nessler's reagent, colorimetric method
Nitrate Nitrogen	Cadmium reduction method
Kjeldahl Nitrogen	Kjeldahl decomposition, titration
Bacteria	Agar plate (pH7.3 nutrient agar media)
Fungi	Agar plate (pH5.6 PDA media)

farming test areas, and then tested after being mixed together. In the soil pile test areas, samples were collected from depths of approximately 10, 30, and 70 cm from a total of 20 locations on both sides of each pile, and tested after being thoroughly mixed.

Plant Toxicity Experiment. Based on laboratory scale experiments, a plant toxicity field experiment was carried out over a period of two seasons using the four kinds of 15-mo bioremediated soil, un-treated contaminated soil and clean desert soil to confirm the effectiveness of soil restoration through bioremediation. Yard 5 x 5 x 3 m, were prepared, then 22 g/m² of alfalfa and 5 g/m² of Bermuda grass were seeded and trimmed once a month. Irrigation and fertilization were performed appropriately. The first crop was tested for plant uptake of toxic materials by measuring EPA 16 PAHs, heavy metals (nickel, vanadium, and lead) and sulphur.

RESULTS AND DISCUSSION

Degradation Behavior of Oil Components. The changes over time in TPH values are shown in FIGURE 2. TPH values decreased rapidly during the first four months and then more gradually over the subsequent months. The degradation rate in the landfarming test sites was higher than in either of the soil-pile test sites during the initial months. After 15 mo, over 80% of the total TPH in the landfarming test sites and over 75% in the soil piles sites had been degraded. The fractional amounts of TEM components are shown in FIGURE 3. Prior to the experiment, the oil consisted of roughly equal thirds of saturated aliphatic compounds, aromatic compounds, and other substances. As is clearly shown in FIGURE 3, the saturated aliphatic and aromatic compounds were broken down, but the resin and asphaltene components were barely degraded. In the landfarming test sites, approximately 80% of the aliphatics and between 40 and 50% of the aromatics had been degraded after 9 mo, and these rates increased to over 90 and 60%, respectively, after 15 mo. In the two soil-pile test sites (not presented) the degradation of aromatics tended to be somewhat

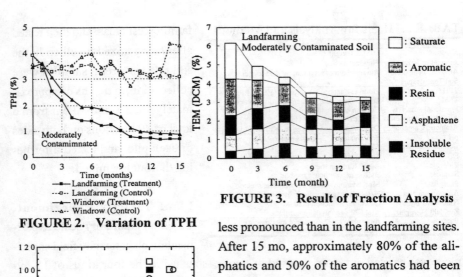

FIGURE 2. Variation of TPH

FIGURE 3. Result of Fraction Analysis

FIGURE 4. The Ratio of each PAH's in Treated Soil Against in Control Soil

less pronounced than in the landfarming sites. After 15 mo, approximately 80% of the aliphatics and 50% of the aromatics had been broken down.

From the GC analysis, the peak values, especially for saturated aliphatics, which have low carbon numbers, fell drastically in the test sites, indicating that progressive degradation was taking place. The ratios of each PAH in the treated soil against the control soil based on HPLC analysis of the 16 types of toxic aromatic compounds designated by EPA are shown in FIGURE 4. This shows that more than 80% of 3-ring PAHs were degraded but 4 to 5-ring PAHs were not degraded after 6 mo of treatment, but 50% of the 4 to 5-ring PAHs were degraded after 12 mo treatment.

Soil-Water Balance. In all three test methods, the moisture content was maintained at between 8 and 10% through watering. The cumulative amount of water used per unit volume of soil in each test method is shown in FIGURE 5. A comparison of the three test methods shows that the landfarming method required more than three times as much water per unit volume of soil to maintain the same moisture content as the windrow and static-pile methods.

Comparison of the Three Methods. Of the three methods tested, the landfarming

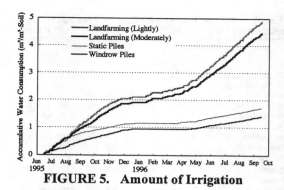

FIGURE 5. Amount of Irrigation

method achieved in the highest decomposition rate and resulted in the greatest reduction in bioremediation time. This technology is also highly appropriate in in-situ applications. However, this method requires much more water to maintain a proper moisture content than the other two methods. The results showed that the windrow-piles method can be more expensive than the others due to the costs incurred in the turning over of the soil with a front loader.

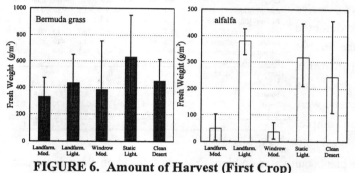

FIGURE 6. Amount of Harvest (First Crop)

Costs are lower with the static pile method, but the application of this method to moderately contaminated soil requires further investigation.

Plant Toxicity. Each experimental plot was divided into four area and investigated. FIGURE 6 shows the average amount harvested in the first crop and also scatterings. The crop of Bermuda grass using treated soil is similar to that using clean desert soil. The crop using treated lightly contaminated soil was larger than that using treated moderately contaminated soil. Thus, the extent of soil remediation seemed to affect the crop. On the other hand, the alfalfa crop grown on treated moderately contaminated soil was very poor compared to that using clean desert soil. That seems to be because of the high salt content of the treated moderately contaminated soil. The harvest of the second crop using treated soil is not shown, but for both Bermuda grass and alfalfa, the crops using any treated soil were better than those using clean desert soil. No germination or growth was observed in the untreated soil plots.

The soil characteristics of first crop were compared with those of the second crop. Then salt content of the soil had reduced remarkably, but the oil content had not reduced as much. It is thought that the salt has a stronger effect on plant growth. It is concluded that oil contaminated soil can be treated well using bioremediation technology, so the treated soil had been sufficiently restored for vegetation.

As shown in TABLE 4, most PAHs were not detected at all. Most of the detected PAHs were under the limit of detection. Fluoranthene, benzo[a]anthracene, and indeno(1,2,3,-cd)pyrene were detected in the plant samples regularly, but there were no difference between the plants cultivated on bioremediated soil and on clean desert soil. PAHs concentration in the plants were lower than that in the soil, so there was no accumulation in the plants. Heavy metals concentration in the plants were same or lower than that in the soil and there was no difference in concentration between plants cultivated on bioremediated soil and on clean desert soil. The sulphur concentration in the plants was about 30% greater than that in the soil. The plants cultivated on the bioremediated soil and on clean desert soil had the same tendency. From the above-mentioned discussion, it was clear that the uptake of PAHs, sulfur and heavy metals by the plants was negligibly small.

TABLE 4. The Concentration of PAHs, Heavy Metals and Sulphur in the Soil and the Test Plants

Item	Landfarming Moderately Contaminated			Landfarming Lightly Contaminated			Clean Desert Soil		
	Soil	Bg	Aa	Soil	Bg	Aa	Soil	Bg	Aa
PAH (µg/kg)									
Naphthalene	<100	n.d.	n.d.	<100	n.d.	n.d.	<100	n.d.	n.d.
Acenaphtylene	<100	n.d.	n.d.	<100	n.d.	n.d.	<100	n.d.	n.d.
Acenaphthene	<100	n.d.	n.d.	<100	n.d.	n.d.	<100	n.d.	n.d.
Fluorene	<100	n.d.	n.d.	<100	n.d.	n.d.	<100	n.d.	n.d.
Phenanthrene	<100	n.d.	n.d.	<100	n.d.	n.d.	<100	n.d.	n.d.
Anthracene	<100	n.d.	n.d.	<100	n.d.	n.d.	<100	n.d.	n.d.
Fluoranthene	<10	17	14	<10	17	<5.2	<10	7.7	17
Pyrene	<100	n.d.	n.d.	<100	n.d.	n.d.	<100	n.d.	n.d.
Benzo[a]anthracene	<10	12	11	<10	10	4.5	<10	4.9	7.6
Chrysene	<100	n.d.	n.d.	<100	n.d.	n.d.	<100	n.d.	n.d.
Benzo[b]fluoranthene	<10	2.3	5.6	<10	4.2	<1.5	<10	1.6	2.2
Benzo[k]fluranthene	<10	0.9	1.4	<10	1.1	1.3	<10	<0.8	2
Benzo[a]pyrene	<10	1.1	1	<10	0.2	<0.7	<10	<0.7	<0.7
Indeno(1,2,3-cd)pyrene	40	1.5	<1.0	20	1	<1.0	<10	<1.0	<1.2
Dibenzo(a,h)anthracene	<10	<0.6	<0.6	<10	<0.7	<0.1	<10	2.6	<0.9
Benzo(ghi)perylene	130	3	<2.2	60	2.6	<2.2	<10	<2.2	<2.5
Heavy metals(mg/kg)									
Vanadium	21	0.49	1.2	18.4	0.83	0.88	20.2	0.65	0.91
Nickel	22	0.98	1.9	20.2	1.3	1.7	22	1.8	3.2
Lead	<20	0.1	0.35	<20	0.2	0.35	<20	0.2	0.25
Copper	6.1	5.5	7	8.6	7.5	7.5	4	4	7
Sulphur(%)	0.23	0.41	0.4	0.14	0.48	0.43	0.21	0.4	0.31

Bg, Bermuda grass, Aa, alfalfa
n.d. not determined

ACKNOWLEDGEMENTS

We are grateful to the Petroleum Energy Center, Japan, and the Kuwait Institute for Scientific Research, Kuwait, for their financial support, and to the Kuwait Oil Company for their cooperation and valuable contribution to the project. We would like to express our sincere appreciation to Professor Matsumoto and Professor Oyaizu of the Department of Applied Biological Chemistry, the University of Tokyo, Professor Omori of the Biotechnology Research Center, the University of Tokyo, and Mr. Makishima of the Analytical Research Center, Japan Energy Corporation for their generous guidance and analysis of oil compounds in this research.

SOIL WASHING AND BIOSLURRY TREATMENT - CLEANUP OF A CLOSED INDUSTRIAL SITE

Gunter H. Brox (TEKNO ASSOCIATES, Salt Lake City, Utah)
William R. Mahaffay (Pelorus Corporation, Evergreen, Colorado)
Bruce S. Yare (SOLUTIA Inc, St. Louis, Missouri)

ABSTRACT: During 1996, SOLUTIA's Everett property near Boston, Massachusetts was remediated in accordance with the State of Massachusetts Contingency Plan (MCP) requirements prior to a reuse of the site as a 700,000 square feet shopping center. 6,000 cubic yard of soil contaminated with bis(2-ethyl-hexyl) phalate (BEHP), a common plasticizer, were treated using soil washing followed by bioslurry treatment of the separated fine soil fraction. The soil washing system was operated by Alternative Remedial Technologies Inc. (ART), and the bioslurry plant was operated by TEKNO ASSOCIATES. Four bioslurry reactors, operated in a semi-continuous mode, provided optimal process conditions for BEHP degradation during a hydraulic residence time of about 6 days. The reactors were seeded with specialized organisms which had been isolated from the site and grown in fermenters for inoculation at start-up and as needed during the project.

INTRODUCTION

A chemical plant was operated at Everett, Massachusetts near the Malden River from the mid - 1800s to 1992. 82 acres of land were cleaned up during 1996 in accordance with the State of Massachusetts Contingency (MCP) requirements prior to the reuse of the site as a 700,000 square feet shopping center. A wide variety of organic and inorganic chemicals were manufactured at the facility. The paper presented describes the remediation of about 1,500 yd^3 soil contaminated with bis(2-ethyl-hexyl) phalate (BEHP) removed from a former rail car loading facility.

PROJECT PLANNING

Off-site incineration had been considered but was abandoned because of high costs. Bioremediation was selected to treat soils containing bis(2-ethyl-hexyl)phthalate at concentrations above the upper concentration limit (UCL) of 10,000 mg/kg. Contaminated soils were excavated, passed through a 6-inch bar screen and 2-inch vibrating screen, and stockpiled pending further separation. During the treatability study phase, BEHP was found to be concentrated in the silt fraction. Soil washing was used to separate the fine fraction, which then underwent bioremediation in aerobic bioslurry bioreactors. Slurry phase treatment was selected because greater process control leads to higher biodegradation rates and thus to shorter treatment times than those observed in solid phase biodegradation studies. Existing equipment was leased, erected, and at the end of the project dismantled and returned to its owners. Bioremediation treatability studies were performed by BIOREM TECHNOLOGIES INC. of Waterloo, Ontario, Canada. These studies showed that the use of well acclimated BEHP degrading microorganisms isolated from the site, and subcultured for further fermentation and inoculation of the bioslurry reactors, acceptable clean-up levels could be achieved using a hydraulic residence time of about 6 days . To further optimize biodegradation kinetics, the temperature was maintained above 30° C (86° F), pH was controlled within a narrow range of about 6.5 to 7.5 and an inorganic nutrient mixture designed for BEHP degraders.

Microbial Inoculum Development

Samples of BEHP- contaminated soil were obtained from excavated soil stockpiles. The soil samples were suspended in sterile Bushnell Haas liquid media at a ratio of 1:5 (w:v) in 500 mL Erlenmeyer flasks containing 150 mL of slurry. The flasks were additionally supplemented with 1000 ppm of technical grade BEHP, capped with a foam plug and incubated at 22° C on a rotary shaker (100 rpms). After 72 hours, 20 mL of slurry were removed and transferred to a fresh sterile Bushnell Haas flask. BEHP was adsorbed onto silica gel, homogenized, and subsequently sterilized in a sealed container. An aliquot of the BEHP/silica was added to the culture flask to achieve an effective BEHP level of 2000 mg/L. Culture flasks were incubated as described until visible growth was observed. A third (tertiary) enrichment cycle was performed before culture isolation was performed.

A BEHP selective agar media was developed. Bushnell Haas (BH) media was prepared at a 2-fold concentration, supplemented with a trace metals solution and sterilized. Nobel agar was added (15 g/500 mL) to deionized water and sterilized. After the sterile liquids had tempered they were aseptically mixed, and 100 grams of sterile BEHP/silica gel was added. The solution was poured into sterile Petri dishes and allowed to solidify.

A serial dilution was performed on each of the tertiary enrichments and dilutions plated on the BEHP selective agar. Plates were incubated at 22° C until visible colonies appeared. Five distinct individual isolates of BEHP degrading bacteria were subsequently selected, streaked for purity on the selective media and archived as frozen glycerol stocks until required.

Inoculum Preculturing and Operation of On-site Fermenters

Preparation of quantities of inoculum was performed in five (5) 25-liter continuously stirred tank reactors. Reactors were sterilized using a chlorine bleach rinse. BH media (20 liters) was prepared and filter sterilized through a 0.22 μm Millipore filter. Each reactor was inoculated with a single pure culture isolate, consisting of 1.0 liter of a 48-hour pre-culture grown on BEHP. Each reactor was then supplemented with 500 mg/L of BEHP and aerated using filter sterilized compressed air supplied through a Porex diffuser. The reactors were thoroughly mixed using a shaft mounted 3 inch pitched blade impeller (lift orientation) connected to a 1/27 hp permanent magnet DC motor operating at 500 rpms. BEHP (500 mg/L) was added twice daily for a period of 72 hours. For the final 24 hours of growth the reactors were supplemented with peptone (2.5 g/L) and yeast extract (1.0 g/L) and a final dose of BEHP (500 mg/L).

When a cell density of between 1.0×10^{10} and 5×10^{10} cells/mL was reached, the culture was harvested. Cell harvesting was done by tangential flow ultrafiltration using a Pelicon cassette system (Millipore Corporation). The cells were concentrated by reducing the volume from 20 liters down to 4.0 liters. The concentrate was packed in ice and shipped overnight to the site for inoculation of on-site operated fermenters.

Two types of fermenters were maintained for inoculation of the bioslurry reactors as needed. A set of three 5 gallon plastic pails was operated inside the heated lab trailer. Mixing and aeration were achieved through small customized rubber membrane diffusers. DOP feed and nutrients were added once daily and approximately one-third of the biomass was harvested daily to inoculate one of three 55 gallon fermenters which were operated outside in the plant area. These fermenters were heated using process steam once the outside temperatures dropped below 15 ° C. The inoculum was fed twice daily with commercial grade DOP and nutrients were added on an as needed basis. It was found that no further inoculation of the full size bioslurry reactors was needed after the initial inoculation at the beginning of the project. The bacteria reproduced adequately in the four slurry reactors and cell plate counts of $10^6 - 10^8$/mL were observed.

REMEDIATION OF BEHP CONTAMINATED SOIL

Bioremediation was selected by SOLUTIA to treat soils containing bis(ethyl-hexyl)phthalate (BEHP) at concentrations above the UCL of 10,000 mg/kg. Impacted materials were excavated, classified in the field using a 6-inch bar screen and a 2-inch vibrating screen, and stockpiled pending further particle size separation. Generally, the material excavated consisted of soil, gravel, concrete/brick rubble, wood, burlap sack pieces, and small amounts of asbestos containing materials.

The excavated soil was first separated using a Reed Screen-All into three fractions: greater than 6-inch, less than 2-inch and greater than 2-inch material. Large debris was segregated manually. Material greater than 2-inch underwent soil mechanical and chemical characterization. Impacted material was sent off-site for disposal. Soils suitable for backfill were returned to the excavation from which they originated. Following gross oversized debris separation at the excavation pit, a wet separation process was subsequently used to further separate and classify the less than 2-inch material. A mobile soil wash plant was operated by Alternative Remediation Technologies of Tampa, Florida. The wet separation included a grizzly to separate any oversize trash that had passed the Reed Screen-All, a vibrating screen to separate the gravel from the sand fraction, a hydrocyclone to separate the fines (silt and clay) from the sand, and a filter press to dewater the fines. Depending on the material, multiple passes were used to complete the separation.

This separation resulted in the generation of three soil streams; gravel (2 inches to 2 mm), sand (2 mm to 0.064 mm) and filter cake (silt/clay < 0.0.64 mm). One soil sample from each of the gravel and sand fractions was collected for every 100 yd³ generated for chemical analysis. The filter cake from the filter press was temporarily stored in the bioremediation soil storage bin to await treatment in the reactors. The soil pile was divided into a grid system and 75 core samples were taken and analyzed for their respective BEHP concentrations. The material was then divided into three soil piles with low (4,000 - 6,000 mg/kg), medium (6,000 - 10,000 mg/kg) and high (> 10,000 mg/kg) BEHP concentrations. (Note: All analytical results are reported on a dry soil basis). The most highly contaminated material was processed first.

Bioslurry Treatment Train

The bioslurry treatment train is illustrated in Figure 1. It comprised a conveyor, a 1,800-gallon slurry mixer, four 2,800-gallon slurry reactors, one storage tank/thickener, several slurry transfer pumps, two low pressure blowers (13 psi), one high pressure compressor (125 psi), and one plate and frame filter press with 36 plates.

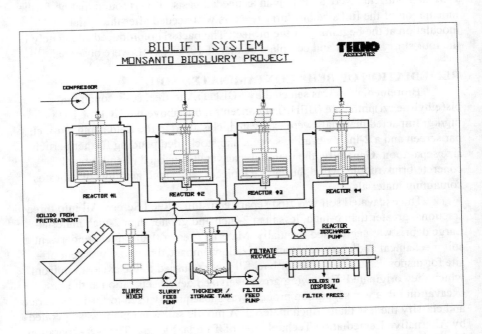

FIGURE 1 BIOSLURRY TREATMENT TRAIN

Material was transported from the storage bin to the conveyor with a Bobcat. The slurry mixer was used to re-slurry the contaminated filter cake to a slurry concentration of about 60 wt%. Prior to transfer, the "feed" to each reactor was supplemented with the pre-weight nutrient mix and the pH adjusted to slightly above 7 using dry sodium hydroxide. The major nutrients added comprised: di-potassium phosphate, mono-potassium phosphate, manganese sulfate, sodium nitrate, ammonium chloride, magnesium sulfate, and several trace minerals. Weekly, each reactor was analyzed for nitrate, ammonium and phosphate and adjustments were made whenever necessary to each respective reactor. From the slurry mixer, the feed was pumped to each of the four reactors using a 2-inch air- driven diaphragm pump. The four reactors were equipped with rakes and rubber diffusers for aeration and mixing. Each reactor also included a mechanical foam breaker which operated only during the start-up phase (after inoculation) when foaming was a major nuisance. Thereafter, foaming was not a problem. In fact, the onset of foaming was determined to be a very reliable biological indicator, that the microorganisms were running out of BEHP substrate.

The reactors were operated in a semi-continuous mode, i.e. every day a portion of the slurry was withdrawn and replaced with fresh feed material. This operation took place only after the slurry was analyzed for the residual BEHP in an on-site laboratory. From the reactor, the slurry was pumped to a storage tank/thickener from where it was pumped to a 48-inch filter press with 36 recessed polypropylene plates. The filter cake was discharged from the filterpress, transported to a holding area and stabilized to reduce leaching of metals. A total of 420 yd^3 of filter cake material was biologically treated during a five-month time period. BEHP concentrations in the feed ranged from 4,000 mg/kg to over 15,000 mg/kg and were reduced to under 3,000 mg/kg.

OPERATIONAL PARAMETERS AND CHEMICAL ANALYSES

The project was staffed with a chemist dedicated to daily GC/MS analysis of the feed to the reactors and the discharged material from the reactors. Once a week, he performed nutrient analyses. BEHP was analyzed from a filter cake sample obtained from vacuum assisted dewatering in a Buttner funnel. EPA Method 3550A (medium/high concentration range) was used to extract BEHP from the filter solids. The extract was then analyzed on a GC/MS following a modified EPA Method 8270 (semi-volatile organic compound analysis of soil samples by gas chromatography/mass spectrometry). The target compound BEHP was identified by comparison of retention time and the mass spectrum of the applicable standards and quantified using a daily calibration standard. As a surrogate, o-terphenyl, and as internal standard, chrysene-d$_{10}$, were used.

A Hach kit and their recommended method for nutrient analysis in a soil matrix was used to analyze nutrient concentrations in the reactors.

To operate the reactors as efficiently as possible the following parameters were monitored on a daily basis:

- **Slurry Temperature**. The objective was to maintain high biodegradation rates. Since BEHP degradation generates heat (exothermic reaction) the slurry temperature remained above 86° F for most of the summer. During the cold season, the slurry was heated through steam injection during the day. Continuous steam injection also throughout the night would have led to too high a temperature.

- **pH.** During the treatability study it had been found that the BEHP degraders worked best in a narrow pH range of 7.0 to 7.5. Since the microbiological degradation process produces phthalate acid, the system had to be buffered to have sufficient alkalinity. Sodium hydroxide was added to the feed slurry in such quantities as to maintain a pH of about 7.0 in the reactors.

- **Dissolved Oxygen (D.O.) Concentration**. A dissolved oxygen concentration above 2 mg/L was targeted to avoid any limitation to the biodegradation kinetics. Air was supplied through 12 rubber membrane diffusers mounted on the rotating rake arms which also contributed to the mixing of the slurry.

- **Oxygen Uptake Rate (OUR)**. The oxygen uptake rate was monitored daily to gain an early insight into any microbiological problems that might develop. This was found to be a very good indicator parameter. Whenever the OUR dropped below 20 mg/L/hr the respective reactor was reinoculated from the fermenter and also slurry from a "more bio-active" reactor was transferred..

- **Slurry Solids**. The reactor type used is a low energy type. It functions best at solids concentrations above 40 wt%. This creates the necessary viscosity to keep coarser particles from settling out too fast which can create a "sanding out problem". Thanks to the rake/airlift combination, some very coarse sand and small gravel could be handled.

- **BEHP Concentration.** The BEHP concentration in all four reactors was analyzed daily except on the weekends. Slurry was only withdrawn when the BEHP concentration had dropped below the 3,000 mg/kg treatment target. Average hydraulic residence time in the reactors was 6 days.

A total of about 420 yd³ of soil fines were treated during a 4-month period. BEHP concentrations in the feed ranged from 4,000 to over 15,000 mg/kg. Treated soil BEHP concentrations were less than the 3,000 mg/kg treatment target.

REFERENCE

Ribbons, D.W., P. Keyser, D.A. Kunz, B.F. Taylor. Microbial Degradation of Phthalates. In: Microbial Degradation of Organic Compounds. (D.T. Gibson, ed.) Marcel Dekker, N.Y. 1984 p. 371.

COMBINED CHEMICAL AND BIOLOGICAL TREATMENT OF MIXED CONTAMINATED SOILS IN SLURRY REACTORS

Koning, M., Lüth, J.-C., Reifenstuhl, R., Hintze, H., Feitkenhauer, H., Stegmann, R.
(Technical University of Hamburg-Harburg, Germany)

ABSTRACT: Comparing and balancing lab-scale investigations were performed to determine if an intermediate ozonation and the use of thermophilic microorganisms are able to induce an advanced biological treatment of mixed-contaminated fine-particle materials in slurry reactor systems. Through the specific effect of an intermediate ozonation PAH (polycyclic aromatic hydrocarbons, particularly higher condensed PAH which are difficult to degrade biologically), significantly could be reduced respectively transformed into water soluble components and thereby made available for a further biological degradation. Contaminant degrading microorganisms were added to enhance the degradation of the dissolved organic contaminants. The use of the presented thermophilic microorganisms allowed a degradation of aliphatic and aromatic hydrocarbons at temperatures of 60-70°C.

INTRODUCTION

In Germany, it is common practice that the fine-particle materials resulting from soil washing, which are often highly contaminated, are landfilled or thermally treated. Biological suspension processes offer an alternative to these procedures because they can significantly reduce the organic contaminant load of the fine particles. However, for the reuse of fine-particle materials advanced treatment processes are necessary.

Within the framework of the Research Center SFB 188 'Treatment of Contaminated Soils', the Department of Waste Management at the Technical University of Hamburg-Harburg conducted comparative and balancing bench-scale investigations into the contaminant degradation during biological treatment of mixed-contaminated fine-particle suspensions containing total petroleum hydrocarbons, PAH and heavy metals (Stegmann, 1998). The aim of the investigations was a) to determine whether the use of thermophilic microorganisms enables biodegradation of contaminants at temperatures of 60 to 70°C, and b) to investigate whether it is possible to induce a further biological contaminant degradation by applying the process combination biological treatment – ozonation – biological treatment.

MATERIALS AND METHODS

The highly contaminated fine-particle material treated in the reactor experiments originated from a processing plant for contaminated railway ballast (Tiefel et al., 1994). The clayey fine-particle material had a maximum particle size of about 10μm and had fairly high contents of organic and inorganic contamination (Table 1).

TABLE 1: Origin material contents [mg/kg dry matter]

Organic contaminants		Heavy metals				Carbon and Nutrients	
TPH	13890	AS	65	Hg	3	TC	145900
EPA-PAH	320	Cd	5	Mn	1266	TIC	4200
		Cr	168	Ni	125	TOC	141700
		Cu	482	Pb	1294	TKN	3200
		Fe	224000	Zn	706	P	1600

PAH: polycyclic aromatic hydrocarbons, TPH: total petroleum hydrocarbons

　　　　The biological treatment of the fine-particle suspensions was performed in mixing reactors with a volume of 12.0 (1st biological treatment) and 4.3 liter (2nd biological treatment), respectively, at different temperatures (20, 40, 60 and 70°C). Figure 1 shows the scheme of the suspension reactors as well as the related data recording and control systems (Koning et al., 1998).

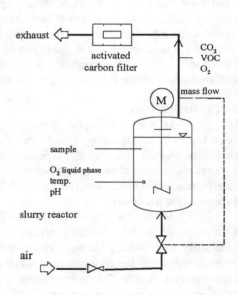

FIGURE 1.　Scheme of the test systems

　　　　To initiate a further biological degradation of contaminants an intermediate ozonation of the suspensions was performed between the first and second biological treatment step. The intermediate ozonation was executed in a 460 l spray reactor at room temperature (Figure 2). The average ozone consumption was $0.12 \ gO_3/gTOC$. During the ozonation, care was taken that the biomass present in the suspension was not completely destroyed by chemical oxidation so that the second biological step could be performed without re-inoculating the fine particle suspension with the reactor setups at 20°C and 40°C. Thermophilic mixed cultures, externally spread onto cellulose carriers, were added to the thermophilic reactor setups (60°C and 70°C) before the first and second biological treatment stage. Table 3 shows the reactor setups chosen for the investigation.

TABLE 2. Reactor setups

Temperature	Ozonation	Additives
20°C	+ after day 29	-
40°C	+ after day 29	-
60°C	+ after day 29	+ thermophilic microorganisms
70°C	+ after day 29	+ thermophilic microorganisms

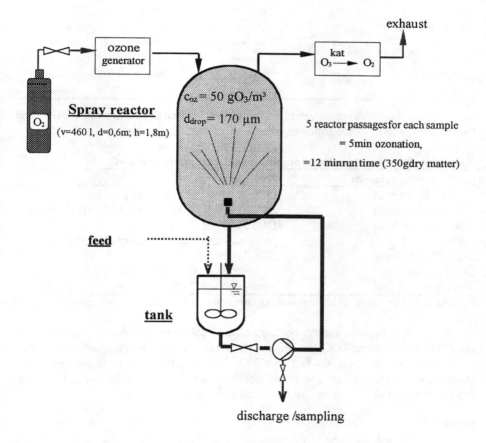

c_{oz}: ozone concentration, d_{drop}: drop diameter

FIGURE 2. Spray reactor for ozonation (Hintze, 1998)

TABLE 3. Used thermophilic microorganisms (Feitkenhauer, 1998)

Bacterial strain	Optimum temperature	Substrates	Culture carbon source	Origin
ISIIa Bacillus thermoleovorans ssp. Hamburg 1	65°C	Aliphatics C_{10}-C_{40} Phenanthrene Pyrene	Eicosane	Hot springs, Iceland
M1 Bacillus spec.	65°C	Aliphatics	Eicosane	Industrial waste water
M2 Bacillus spec.	65°C	Aliphatics	Eicosane	Industrial waste water
M3 Bacillus spec.	65°C	Aliphatics 1-Chlorhexadecane	Eicosane	Industrial waste water
D5 Bacillus spec.	65°C	Synth. diesel fuel	Eicosane	Hot springs, Iceland
Comp. Bacillus spec.	65°C	Aliphatics	Eicosane	Org. compost
GZ	70°C	Aliphatics, (PAH)	Eicosane	Org. compost
Bacillus thermoleovorans	65°C	Aliphatics C_{12}-C_{20}	Eicosane	Type Strain DSM
HH2 Bacillus thermoleovorans ssp. Hamburg 2	60°C	Naphthalene Benzene Toluene	Naphthalene	Cont. Compost
A2 Bacillus spec.	65°C	Phenol	Phenol	Hot springs, Iceland
B5 Bacillus spec.	60°C		Benzoic acid	Compost

To estimate the turnover of contaminants, carbon balances were created and the carbon contents of the following components were determined:

TABLE 4. Measured carbon compartments and analytical methods

Carbon compartment	Analytical method
TOC (solid matter)	DIN 38409 H3-1 (modified)
TPH (solid matter)	DIN 38409 H18 (modified), GC-FID (UBA)
polar hydrocarbons	DIN 38409 H18 (modified)
biomass	protein determination (Lowry et al., 1951)
DOC (liquid phase)	DIN 38409 H3-1
CO_2 (exhaust gas)	on-line: IR ADC DB2E
VOC (exhaust gas)	discontinuous: TOC-analyzer

TOC: total organic carbon, DOC: dissolved organic carbon, VOC: volatile organic carbon

The organic carbon contents (Corg, solid matter) were calculated from the difference between TOC and the analytical captured carbon compartments (TPH, polar hydrocarbons, biomass and DOC). In addition, the pH-values, temperatures, O_2-contents in exhaust air and liquid phase as well as the 16 EPA-PAH's in the examined suspension samples (EPA method 610) were determined.

RESULTS

To estimate the biological activity in the suspension reactors, the actual CO_2-production was used as an indicator. In all reactor setups the CO_2-production decreased below 0.6 mgC*kg dry matter^{-1}*h^{-1} (max. 13.0 mgC*kg dry matter^{-1}*h^{-1}) after day 29 and it was, therefore, assumed that the primary degradation processes were completed at this stage. A comparison of the carbon components at the beginning and at the end of the first biological treatment (figure 3 shows the carbon balance of the 20°C experiment) shows that the carbon contents of the TPH, polar hydrocarbons and biomass decreased in time and were transformed into DOC and CO_2. Larger quantities of the TOC were detached from the soil particles and collected at the reactor walls (material deposit). In contrast, the amount of volatile organic Carbon (VOC) stripped out of the suspension was negligibly small.

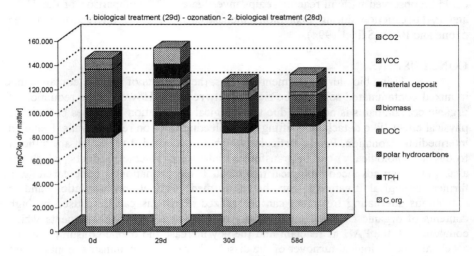

FIGURE 3. Carbon balance for the complete treatment process at 20°C

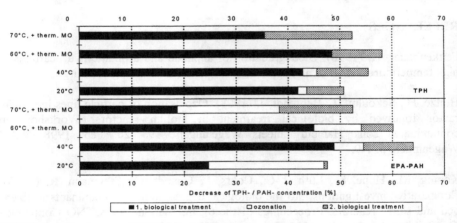

**FIGURE 4. Decrease of TPH- and PAH concentrations
(Biological treatment / ozonation / biological treatment)**

To enhance a further biological treatment, the suspensions were ozonised as previously described. The material deposit was not transferred into the spray reactor for ozonation and is therefore not included in the carbon balance of the second biological treatment step (day 30 - day 58). The ozonation of the suspension primarily decreased the organic contents of the solid matter, Corg (including PAH), and increased the content of DOC. Together with fairly large parts of the TPH and polar hydrocarbons, large quantities of the DOC were mineralized to CO_2 in the second biological treatment step.

The effect of each single treatment step on the reduction of TPH and PAH is shown in figure 4. A considerable decrease of the TPH and PAH contents could be observed with all reactor setups investigated. The comparison of the TPH- and PAH-reduction during ozonation emphasizes the specific interaction between ozone and PAH (Seidel,1994).

CONCLUSIONS

Using bio-slurry-treatment-systems, the contents of organic contaminants in mixed-contaminated soils can be significantly reduced. However, the reduction of organic contaminants is not only due to biological degradation processes but also to physical effects like detaching, shifting and collecting resp. on reactor walls. Using an intermediate ozonation, PAH (particularly higher condensed PAH which are difficult to degrade biologically) can be transformed into water soluble products, which are available for an advanced biological treatment. If the ozonation is done carefully, a further biological treatment can be performed without re-inoculation and a continuous alternating treatment can be realized. Problems can be caused by high contents of organic carbon (Corg) and by reduced heavy metal components which compete with the PAH as reactant and thereby increase the ozone consumption. To accelerate the biological turnover of the dissolved organic contaminants, contaminant degrading microorganisms can be added. In this context, the presented thermophilic microorganisms allow a biological degradation of aliphatic and aromatic hydrocarbons at temperatures of 60 -70°C.

REFERENCES

Feitkenhauer, H. (1998): Biodegradation of aliphatic and aromatic hydrocarbons at high temperatures: kinetics and applications, Diss., TU Hamburg-Harburg.

Hintze, H., Sekoulov, I., Behrendt, J. (1998): Ozonisation of particle fines in a spray ractor followed by biological treatment in biofilm reactors, Workshop on contaminated soils 'Hamburg meets Wageningen', 19.- 20. Feb. 1998, WAU Wageningen, p. 7-9.

Koning, M., Hupe, K., Lüth, J.-C., Cohrs, I., Quandt, C., Stegmann, R. (1998): Comparative investigations into the biological degradation of contaminants in fixed-bed and slurry reactors, Proceedings of the 6th International FZK/TNO Conference 'ConSoil '98', 17.-21. May 1998, Edinburgh, UK, p. 1091-1092.

Seidel, J.-P. (1994): Über die Anwendung von Ozon zum oxidativen Abbau polyzyklischer aromatischer Kohlenwasserstoffe in Böden, Diss., TH Karlsruhe.

Stegmann, R. (1998): DFG-Research Centre SFB 188 'Treatment of contaminated soils' - Summary -, Technical University of Hamburg-Harburg.

Tiefel, H., Lohmann, G., Donhauser, F. (1994): A processing plant for contaminated railway ballast, Aufbereitungs-Technik 35, Nr. 10, p. 515-523.

INFLUENCE OF OXYGEN ON THE DEGRADATION OF TPH-CONTAMINATED SOILS

Karsten Hupe (Consultants for Waste Managment, Hamburg, Germany)
Joern Heerenklage (TUHH, Hamburg, Germany)
Rainer Stegmann (TUHH, Hamburg, Germany)

ABSTRACT: The influence of oxygen on the biodegradation of diesel fuel in unsaturated soil/compost mixtures at 30 °C was analyzed over a period of 7 weeks. Firstly, a wide range from 0 to 80 vol. % O_2 was investigated. Afterwards, the range below 5 vol. % O_2 was examined more closely. Regarding the whole test period of 7 weeks no significant influence of the oxygen content could be observed above 1 vol. % O_2. Anaerobic conditions should be avoided for the degradation of diesel fuel. Furthermore, a model was developed to estimate the total mineralization as a function of the oxygen content.

INTRODUCTION

For controlling the aeration rates in biopiles or in bioreactors, the optimal oxygen concentration range in the inlet air and especially in the soil material have to be measured in bench scale or pilot scale test series. The literature contains only a few turnover balancing examinations about possible effects of different oxygen concentrations in water unsaturated soil materials.

For water saturated systems (sediments, suspensions), a series of results about the influence of the dissolved oxygen concentration on the microbial turnover of hydrocarbons is available: Michaelsen et al. (1992) ascertained a significant influence of the dissolved oxygen concentration on the turnover of hexadecane in a sediment/sea water suspension when falling below 1% pO_2 (oxygen partial pressure). Within 190 hours no hexadecane turnover could be measured under anaerobic conditions. At 0.4% pO_2 a turnover and a mineralization of hexadecane could be ascertained but with a longer lag-phase.

A series of laboratory tests are described in which the influence of different oxygen concentrations (0%-80%) in the inlet air on the microbial conversion of diesel fuel in soil/compost mixtures at 30°C were examined.

MATERIAL AND METHODS

Test units. Reactor systems (volume 3 liter) are applied to simulate the conditions in an aerated windrow or in large-scale reactors. Static bioreactors called fixed-bed bioreactors including on-line recording and control of the measured values were used for the investigations. In these systems the volatilization can be measured accurately. The investigations took place under defined temperature conditions in climatic chambers at 30 °C (detailed description: Hupe et al., 1997).

Three test series (OX1-3) were carried out to investigate the influence of oxygen content on the biodegradation of diesel fuel. In OX2 and OX3 the CO_2 concentrations in the exhaust air stream were analysed by an on-line IR (infrared) photo-spectrometer and the VOC (volatile organic carbon) concentrations were measured by an on-line FID (flame ionization detector). Further gas samples could be taken with a syringe through a septum from the upper part of the reactor and could be injected then into a GC (gas chromatograph). This system was also used for the discontinuous measuring of the CO_2 concentration and the VOC concentration in OX1.

Soil preparation. A slightly loamy sand (sieved to ≤ 2 mm) was contaminated artificially with 1% (% of dry weight of soil) diesel fuel. Mature biocompost (sieved to ≤ 4 mm; degree of maturity: V) was used as soil supplement. The water contents of the soil/compost mixtures were adjusted with distilled water to 55% resp. 60% of the maximum water-holding capacity (WC_{max}).

　　　　Aeration of the soil materials was carried out by means of synthetic air (O_2/N_2 mixtures). While the aeration rate for test series OX1 was adjusted to a fixed value of 2 l/h for the whole test period of 48 hours, the aeration rates in test series OX2 and OX3 were regulated following the expected oxygen consumption determined in preliminary tests and then converted into an aeration rate for each test serie respective for each oxygen concentration with a security factor of 3. In Table 1 the test set-ups are described.

TABLE 1. Parameters of test series OX1-3

Test series	OX1	OX2	OX3
Soil material [g dwt]	1420	1045	800
Diesel fuel [g/kg dwt]	10	10	10
Compost [g dwt]	360	105	80
Soil/compost ratio	4:1	10:1	10:1
pH	6.9	6.7	6.7
WC_{max} [g H_2O/100 g dwt]	55.9	49.2	45.6
Water content [% of WC_{max}]	60	55	55
Temperature [°C]	30	30	30
Oxygen content [vol. %]	0; 5; 10; 21; 40; 80	1; 2; 3; 4; 5; 21	0; 0.1; 0.2; 1; 2; 5; 21
Gas analysis	Discontinuous	on line	on line
Flow rate [L/h]	2	varying #	varying #
Test period [d]	48	49	7

dwt: dry weight; WC_{max}: maximum water-holding capacity; #: depending on microbial activity

Carbon Balance of Oil Degradation. During the investigations the following carbon balancing parameters were measured: hydrocarbon content in the soil, biomass, CO_2 and VOC (detail described in Hupe, 1998).

　　　　The difference between CO_2 production of the contaminated samples and the uncontaminated samples which were referred to the carbon content of the initial contamination is called mineralization in the following (stated as % C of origin).

The decrease of pollutants as a function of mineralization was described by a model of 1st order. To increase the comparability of the measured results, the balancing sizes were referred to the recovery rate of the pollutants in the origin. This means that the recovery rate of the origin was put at 100% in the test series.

For a summarizing examination of the influence of the oxygen content on the pollutant's turnover, the mineralization rates of test series OX1, OX2 and OX3 were compared at different moments of the pollutant's turnover. The cumulative mineralization of the different test series was applied after 7 days of the test period as well as at the end of the test (after 7 weeks) against the oxygen content and was evaluated analogously to the Monod kinetics formulation (equation 1).

$$\Delta C_{CO2} = \frac{s_1 \cdot C_{O2}}{s_2 + C_{O2}} \tag{1}$$

with:

ΔC_{CO2} cumulative mineralization [% C of origin]
C_{O2} oxygen content in the aeration stream [vol. % O_2]
s_1, s_2 rate constants [% C of origin], [vol. % O_2]

INVESTIGATIONS AND RESULTS

Decrease of Hydrocarbon and Mineralization. In OX1 the diesel fuel contamination decreases in an exponential way from 100% contamination at the beginning to approx. 15% [% C of origin] independend on the oxygen content. Only the test set-up with an O_2 content of 0% showed a low decrease to 80% of origin. The degree of mineralization of the contamination shows a consistent value of approx. 55% [% C of origin] for the test set-ups with 21% O_2, 10% O_2, 40% O_2 and 80% O_2 after 48 test days. The reactor aerated with 5% O_2 showed -at the end of the test with the same remaining contamination - a mineralization which was approx. 15% higher. The mineralization for the reactor aerated with pure nitrogen amounted to 2% at the end. The m_1-factors calculated according to exponential equation are close together between $m_1=0.028$ to $m_1=0.033$ for the reactors aerated with oxygen. The calculated factor rises with increasing oxygen content in the inlet air. Only the reactor aerated with pure nitrogen shows a factor of $m_1=0.117$.

Due to the lower compost addition of 10% dry weight in OX2, the mineralization rates are lower compared to those of series OX1 with 20 weight-%. After the end of the test, the cumulative CO_2 production amounted to 50% for the reactor aerated with 21% O_2, to 49% for the reactors aerated with 5% and 3% O_2 and to 44% for the reactors aerated with 4% and 2% O_2. The reactor aerated with 1% O_2 shows a mineralization rate of 41% after 49 days. The contamination decreased during the test period to values between 10 and 22%. The m_1-factors calculated according to exponential equation are in a close range and show values between $m_1=0.032$ and $m_1=0.041$. These numbers corresponds to the m_1-factors of series OX1.

Oxygen Content. Figure 1 shows the influence of the oxygen content on the total mineralization. For the evaluation of the oxygen influence, the total mineralization after 7 days and at the end of the test was evaluated. The two curves were determined by equation 1. Apart from series OX1 and OX2, this description also takes into consideration further data of test series OX3 concerning the oxygen content which was carried out following OX1 and OX2.

The ascertained total mineralization for oxygen contents between 0 and 80% in the inlet air in the phase of the maximum degradation rate after 7 days follow the equation $DC_{CO_2}=18.78\% * C_{O_2} / (1.779$ vol. % $+ C_{O_2})$. During the whole test period, the determined mineralization rates follow the equation $DC_{CO_2}=56.71\% * C_{O_2}/(0.23$ vol. % $+ C_{O_2})$. Taking account these test conditions and the varying oxygen contents the above mentioned equations can be used to estimate the total mineralization.

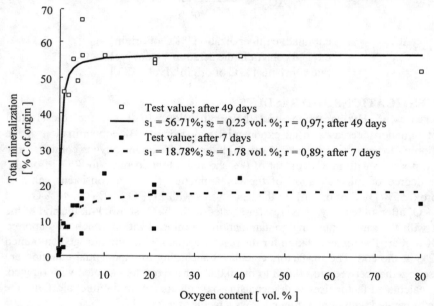

FIGURE 1. Influence of the oxygen content on the total mineralization (s_1 and s_2 analogous equation 1)

DISCUSSION

At the end of the test (7 weeks), no significant influence of the oxygen in the inlet air in the range between 1 - 80 vol. % oxygen on the total oil turnover could be ascertained between the different set-ups of test series OX1 and OX2. However, in the course of the pollutant's conversion an influence could very well be analyzed. This was reflected in the different specific CO_2 production (see Hupe et al., 1998) and the different values of total mineralization at the beginning (after 7 days) of the investigations (see Figure 1). Thus the course of the CO_2 production demonstrated that with a low O_2 concentration the activity maxima was also lower and occurred with delay.

The small influence of the oxygen concentration respecting the whole test period can be explained by the fact that after the first degradation phase of the easily biodegradable compounds, only the less degradable (iso-alkanes etc) and carbon compounds remained in the soil whose conversion is not limited by oxygen. Under anaerobic conditions a mineralization of only 2% and a pollutant decrease of 21% could be measured. These results are comparable to those found out by Michaelsen et al. (1992). They came to the conclusion that no influence on the conversion of hexadecane in sediment-seawater could be detected in the range between 5% and 80% of the oxygen saturation in the water (comment: the 5% set-up of the investigations of Michaelsen et al. (1992) corresponds to about the 1% set-up of test series OX2).

The different values of total mineralization - with nearly the same total pollutant decrease in test series OX1 and OX2 and similar test conditions - concerning water content, temperature, pH value, contamination, soil material and reactor system - were put down to the different quantities of added compost. While 20% dry weight compost were added to the soil material of OX1, only 10% dry weight compost were added to the soil material of OX2. Investigations concerning the influence of the compost content on the pollutant's decrease verify the explanation that the higher the compost content the higher the mineralization and the higher the oil turnover (Hupe et al., 1996). According to that an increasing pollutant mineralization of diesel fuel could be determined but an increased oxygen supply is also necessary.

It could be observed that the equation 1 is useful to estimate the total mineralization as a function of the available oxygen content. More investigations with varing organic contaminations (quality and quantity), soil materials and milieu conditions (e.g. pH, temperature) are necessary in order to validate this equation. In this case, this equation or a modified one could be integrated into a model for the prediction of the biodegradation of organic contaminations with time.

The applied test system has proven to be well suitable to investigate and balance the conversion of the pollutants in detail especially due to the development of the automatic on-line measuring device for the exhaust air. This enabled to measure the second maximum of CO_2 production for the first time. Under discontinuous CO_2 measuring the second maximum could not be detected, since the rhythm of sampling was expanded after the first week of the test series from two samplings per day to one sampling per day and after the fourth week only two samplings per week were taken.

For the biological degradation of diesel fuel in soil it can be concluded that under comparable milieu conditions - as described in this work - there is no limiting influence of the oxygen when the concentration is higher than 1 vol. %, of course anaerobic conditions have to be avoided. The influence of low oxygen contents - especially in the range between 0 and 5 vol. % O_2 - on the biodegradation of hydrocarbons has to be investigated in detail.

An increased oxygen supply is necessary for the biological treatment of recently contaminated soil materials - especially at the beginning of the treatment (approx. 4 weeks). This can be realized by active aeration (see Koning et al., 1999), by reduction of the gas diffusion pathways (reduced height of the windrows or landfarming), by turning and/or by mixing of the windrows. Furthermore, for the bioremediation of soil contaminated with diesel fuel in the windrow process the influence of the oxygen supply decreases the older the contaminated site is. This can be explained by the fact that the well bioavailable compounds of the contaminants disappeared to a great extent in the first weeks so that the oxygen supply is not very important for the remaining contamination. Nonetheless, it has to be taken into consideration that anaerobic conditions have to be avoided.

ACKNOWLEDGEMENT

The investigations represented in this work took place within the scope of the investigation program of the research center SFB 188 of the DFG.

REFERENCES

Hupe, K. 1998. *Optimierung der mikrobiellen Reinigung mineralölkontaminierter Böden in statischen und durchmischten Systemen.* Hamburger Berichte 15, Verlag Abfall Aktuell, Stuttgart, Germany.

Hupe, K., J. Heerenklage, H. Woyczechowski, S. Bollow and R. Stegmann. 1998. „Influence of Oxygen on the Degradation of Diesel Fuel in Soil Bioreactors." *Acta Biotechnol. 18*(2): 109-122.

Hupe, K., J.C. Lueth, J. Heerenklage and R. Stegmann. 1997. „Test systems for balancing and optimizing the biodegradation of contaminated soils: A German perspective." In S.K. Sikdar and R.L. Ivine (Eds.), *Fundamentals and applications Volume 1 "Bioremediation: Principles and practice",* pp. 665-692, Technomic Publishing Co., Inc., Lancaster, Basel.

Hupe, K., J.C. Lueth, J. Heerenklage and R. Stegmann. 1996. „Enhancement of the biological degradation of soils contaminated with oil by the addition of compost." *Acta Biotechnol. 16*(1):19-30.

Koning, M., J. Braukmeier, J.-C. Lueth, N. Viebranz, V. Schultz-Berendt and R. Stegmann. 1999. „Optimization of the Biological Treatment of TPH-Contaminated Soils in Biopiles." In A. Leeson and B. Alleman (Symposium Chairs), Proceedings 5th International Symposium on *In Situ and On-Site Bioremediation,* April 19-22, 1999, San Diego, California.

Michaelsen, M., R. Hulsch, T. Hoepner and L. Berthe-Corti. 1992. „Hexadecane mineralization in oxygen-controlled sediment-seawater cultivations with autochthonous microorganisms." *Appl. Environ. Microbiol. 58:* 3072-3077.

MICROBIAL DEGRADATION OF OIL CONTAMINATED BIOSLUDGE IN A FULL SCALE COMPOST REACTOR

O. Bergersen, T. Briseid, and G. Eidså (SINTEF, Oslo, Norway).
A. Vaa (Vaa Biomiljø AS Sauland, Norway)

ABSTRACT: A continuous composting process has been run in a $23m^3$ horizontal cylinder reactor connected to a biofilter. The material used in the process was oil-contaminated biosludge from a wastewater treatment plant of an oil refinery. The sludge was mixed with pine bark in a ratio of 1:1 and added fertilizer before composting. The high content of available carbon in the biosludge resulted in high temperatures and high O_2 consumption and CO_2 production rates which indicated high microbial activity in the reactor process. Chemical analysis (GC/MS) of the compost showed that the high microbial activity was accompanied by a high reduction of PAH (polyaromatic hydrocarbons). Even heavy degradable hydrocarbons as phytane and hopanes and PAH compounds like pyrene and benzo(a,e)pyrene showed reduction during the process. The concentrations were further reduced during storage after the reactor process. The reduction of oil compounds was accompanied by a reduction in measured toxicity and a reduction of the strong and unpleasant odor of the biosludge.

INTRODUCTION

Historically, land fills or land farming has been used for treatment of organic wastes from petroleum and chemical processing industries. Today composting or incineration is also often used for treatment of the wastes. Land farming is still usually the most cost efficient method. However the microbial degradation of oil sludge by land farming is a slow process with limited degradation of higher polycyclic aromatic hydrocarbons (PAHs) (Wilson and Jones, 1993). Production-scale trials have been done on the decontamination of oil-polluted soil in a rotating bio-reactor at field capacity. (Munkhof, van den, and Veul, 1991) Kristjansson and Stetter (1991) give an overview of thermopile bacteria. Many termophilic hydrocarbon-utilizing bacteria have been isolated from a variety of environmental samples. (Perry, 1985; Premuzic and Lin, 1991). These bacteria have been used for treatment in laboratory reactor at 60°C of sludge containing PAH and degradation of waste from refinery (Castaldi et al., 1995).

The aim of this work has been to study the treatment of oil contaminated biosludge mixed with bark as matrix under aerobic conditions at 50-70°C in a rotating full scale compost reactor.

MATERIAL AND METHODS

Compost reactor system. The volume of the rotating compost reactor was $23m^3$. The effluent air from the reactor ($32m^3$/hr) was treated by a closed biofilter ($2m^3$) mainly filled with compost and spruce and pine bark (reactor and biofilter were constructed by Vaa Biomiljø a.s). The oil-contaminated biosludge from a water treatment plant of an oil refinery (1-2% oil on wet basis and 10-15% dry weight) was mixed with pine bark in a ratio of 1:1 and added fertilizer before composting. The mixture (about $1m^3$ per day) was fed the front end and the transport time through the reactor was 18-20 days. The produced compost was then stored in piles for about 40 days.

Measurements of temperature and concentrations of O_2, CO_2 and H_2S. The temperature was measured and logged every hour in the reactor mass at three different places, both in the front of the reactor, in the middle of the reactor and near the end of the reactor. The volume % of O_2, CO_2, and H_2S and %LEL (% of lower explosion level) were monitored in the effluent ventilated air from the reactor.

Chemical analyses. Chemical analyses were performed on biosludge, bark, composted material and stored compost. Initial concentrations of the mixture of bark and biosludge (1:1) which was used in the process were calculated based on this analysis. The extraction procedure was modified after Intergovernmental Oceanographic Commission, (1982). Total hydrocarbon (THC) was analyzed by GC/FID while polycyclic aromatic hydrocarbons (PAHs) were analyzed by GC/MS. The volatile organic compounds (VOC) in the effluent air from the reactor were trapped on silica and/or charcoal absorbents and analyzed by GC/MS.

Bacterial number. Bacterial suspensions from the samples were prepared by mixing 180 ml mineral solution (Perry. 1985) with 20g samples. The total bacterial numbers were estimated using CFU (NA agar plates, Nutrient Broth, Difco) and oil degrading microorganisms using most probable number technique (MPN). The agar plates and the bottles used for MPN were incubated for 7 days at 28°C (mesophilic microorganisms) and 55°C (termophilic microorganisms) before counting.

Microtox analysis (EC 50 value). The Microtox screening test used (Beckman Microtox TM System) was a bacterial test measuring acute toxic effects. Photoluminescence bacteria (*Photobacterium phosphoreum*) were used as test organisms. Water extracts used as test samples were prepared by mixing 10% wet weight of compost or sludge with water for 3 hours. The suspensions were separated by sentrifugation for 15 min at 2500 rpm and the supernatants used as test samples.

RESULTS AND DISCUSSION
Compost characterization. The pH of the pine bark was about 4,8 and the bio sludge was about 6,0. The pH increased to 7,5 after 10 days of composting and further to 8.5 after 20 days composting in the reactor. However, afterwards it decreased to 7,5 after 40 days storage in piles outdoors. The dry matter increased during the process (table 1). The total amount of termophilic bacteria increased

during the reactor process from about 10^6 to about 10^8 per/g compost but decreased during storage afterwards to about 10^7 per/g.

The measured toxicity of the compost was lower after the process than the toxicity of the biosludge and the toxicity of the pine bark, which was filled into the reactor. Low numbers in table 1 means that only small amounts is needed in the test to obtain a defined toxic effect.

Temperature and concentrations of O_2, CO_2 and H_2S. The average temperatures in the mass in the reactor were 68°C in the front, 67 °C in the middle and 60°C in the end of the reactor. Average volume % of CO_2, and O_2 in the effluent air were 4,0 and 18,6 respectively. Measurement of H_2S showed average values of 3,3 ppm. Maximum values of lower explosion level were 10% to 12%. These maximum values were obtained during the rotation of the reactor.

It was important to measure LEL values continuously during the experimental period in order to avoid explosive concentrations of VOC in the process.

TABLE 1 Characteristics of biosludge, pine bark, compost after 10 and 20 days composting and stored compost after 40 days. [a]

Sample	pH	C/N ratio [b]	Dry Matter (%)	EC 50 [c]	CFU [d]	MPN [e]
Biosludge	6,0	8,3	10	2000	1×10^6	1×10^3
Pine bark	4,8	-	40	4400	2×10^6	4×10^{4l}
Compost after 10 days	7,5	19,0	35	4300	2×10^8	1×10^7
Compost after 20 days	8,5	20,0	37	11300	3×10^8	9×10^4
Stored compost after 40 days	7,5	22,3	43	27300	3×10^7	3×10^3

[a] The data are average values determined on a dry weight basis.

[b] Total C% (dry weight) / Total N % (dry weight)

[c] Microtox values EC 50 measured in ppm / wet weight.

[d] Colony forming units (CFU) growing at 55 C on Na plates.

[e] Most probably number (MPN) of bacteria per g. growing on oil (diesel).

PAH and THC. The bio-sludge used in the experiment contained, 113000 mg THC per kg dry matter. The results show a THC reduction of 61% during the reactor process (20 days) and 71% reduction after 40 days of storage in piles. The reduction of some PAH compounds is shown in table 2 and in fig. 1(A, B and C). Mono and bicyclic aromates were reduced by 92 % during the reactor process and further decreased during storage. The highly resistance compounds chrysene/triphenylene and benzo(a.e.) pyrene showed very high reduction during the process and were reduced by 90 % after 40 days storage. Pyrene was not identified in the stored compost.

TABLE 2. Percentage reduction (by weight) of THC, identified mono and bicyclic aromates, total identified PAH and selected oil components.

		Reduction (%)		
Nr.	Compounds	Compost in the middle (10 days)	Compost out of reactor (20 days)	Compost stored for 40 days
	Total hydrocarbon carbon	21	61	71
1	Mono, bicyclic aromate	66	92	96
2	Total PAH	59	82	94
3	Pyrene	50	86	100
4	Chrysen and Triphenylen	65	74	91
5	Benzo(a,e)pyrene	47	59	87
6	Alkanes	63	83	98
7	Alkanes & acyclic isoprenoides	59	87	97
8	Bicyclonaphalens	62	77	97
9	(n-$C_{18} H_{38}$)	52	89	97
10	Hopanes	74	86	91

A method using the relative ratio between easily degradable and more resistant compounds has been successfully used to investigate changes in spilled oil (Wang et al. 1994). This method is also useful in order to study the degree of oil biodegradation. Different relative ratios are shown in table 3. One should be aware that the oil components in biosludge are concentrated with respect to oil compounds that are difficult to degrade.

TABLE 3 Relative ratio between different compounds.

		Relative ratios		
Compounds	Biosludge	Compost in the middle (10 days)	Compost out of reactor (20 days)	Compost stored for 40 days
Alkanes/acyclic isoprenoides	1,4	1,3	1,0	0,7
(n-$C_{18}H_{38}$)/phytane	0,5	0,2	0,1	0,1
(n-$C_{18}H_{38}$)/hopanes	3,1	1,7	0,1	0,1
Phytane/hopanes	6,5	3,8	0,5	0,8

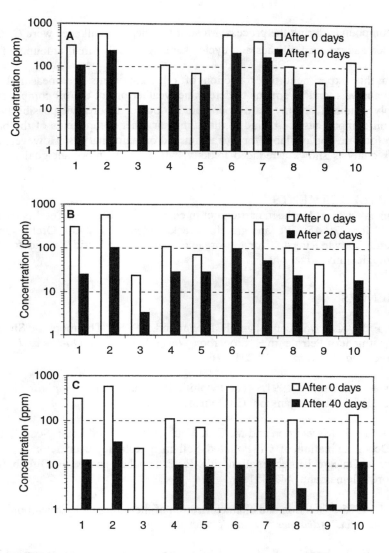

FIGURE 1. Concentration profiles (in ppm) of different PAH compounds (1-10, table 2). Biosludge mixed with bark compared with samples after 10 days composting (A), 20 days composting (B) and 40 days pile storage (C).

More easily degradable compounds are already degraded in the water treatment plant. The results in this work show that also the concentration of very resistant compounds is very much reduced during the process.

The amounts of volatile organic compounds (VOC) in the outlet air were measured to 450–510mg/m^3 or about 5 % of the average amounts of the THC added. Most of the identified VOC was monoaromates like toluene, ethylbenzene and xylenes. This may explain the effective reduction of these low molecular weight compounds during the process. Most of the identified alkanes were C_6-C_{12} compounds as branched alkanes and cycloalkanes with only small amounts of n-alkanes.

In this experiment the reduction of oil components are measured by chemical analysis, and it remains to determine weather or not the oil components are simply oxidized or only partly converted. Another theoretical possibility is that the oil components are bound to the humus or other components of the bark and therefore are not available for extraction and organic analysis. However, the Microtox analysis shows a reduced toxicity, which is an important goal of the process.

ACKNOWLEDGEMENTS

The financial support has been carried out in co-operation between Total Norway, and Vaa Biomiljø AS. We are gratefully acknowledged Frøydis Oreld, Grete Tveten and Oddvar Ringstad for the organic analysis and Turid Røneid for the microbiological investigations.

REFERENCES

Castaldi, F. J. , K. J. Bombaugh, and B. McFarland. 1995. "Thermophilic Slurry-Phase Treatment of Petroleum Hydrocarbon Waste Sludges." *In On-site and In-situ Bioreclamation symp. 3* (8): 231-246.

Kristjansson and Stetter. 1991. "Thermophilic Bacteria." In *Thermophilic Bacteria.*" Ed. J. K. Kristjansson. CRC Press.

Munkhof, Ger P. M. van den and M. F. X Veul, 1991. "Production-Scale Trials on the Decontamination of Oil-Polluted Soil in a Rotating Bioreactor at Field Capacity. In On-Site Bioreclamation, eds. R. E. Hinchee and R. F. Olfenbuttel. Butterworth-Heinemann. 443-451.

Perry, J. J. 1985 "Isolation and characterization of thermophilic hydrocarbon-utilizing bacteria." *Advances in Aquatic Microbiologi 3:*109-139.

Premuzic, E. T., and M. S. Lin. 1991. "Interaction between thermophilic microorganisms and crude oils: recent developments." *Resources, Conservation and Recycling 5*: 277-284.

Wang, Z., M. Fingas, and G. Sergy. 1994. Study of 22 year-old Arrow oil samples using   biomarker compounds by GC/MS. *Environ.Sci. Technol. 28*(9):1733-1746.

Wilson, S. C., and K. C. Jones. 1993. "Bioremediation of Soil contaminated with Polynuclear Aromatic Hydrocarbons (PAHs): A Review." *Environmental Pollution 81*: 229-249.

OPTIMIZATION OF THE BIOLOGICAL TREATMENT OF TPH-CONTAMINATED SOILS IN BIOPILES

Koning, M., Braukmeier, J., Lüth, J.-C., Ruiz-Saucedo, U., Stegmann, R.
(Technical University of Hamburg-Harburg, Germany)

Viebranz, N., Schulz-Berendt, V.
(Umweltschutz-Nord GmbH & Co., Germany)

ABSTRACT: The rate of total petroleum hydrocarbons (TPH) degradation in biopile soils undergoing biological treatment is substantially influenced by the availability of oxygen in the biopiles. Passive aeration (diffusion through the surface of the piles) can only provide sufficient oxygen in depths of less than 2 meters within economical treatment periods (up to 1 year). For biopiles with heights of up to 5 meters (which are more economical due to lower space requirements), active aeration systems are necessary. In biopiles with heights of 3.2 meters, sufficient oxygen could be provided with an oxygen-controlled, discontinuous, high-pressure aeration system. A continuous TPH reduction could be achieved with this system. Lab-scale investigations carried out simultaneously showed that less than 5 % of the original contamination is stripped when this system is applied.

INTRODUCTION

Within the scope of the research at the DFG-Research Centre SFB 188 "Treatment of contaminated soils", balancing investigations on the biological degradation of TPH in contaminated soils were carried out at the Department of Waste Management of the Technical University of Hamburg-Harburg. The investigations were carried out in lab-scale test systems (respirometers, solid-state reactors, slurry reactors) and at technical treatment plants (biopiles, slurry reactors). As a result of these comparative investigations, a pilot plant for the oxygen-controlled high-pressure aeration was developed and currently is being tested in a collaborative project with Umweltschutz Nord GmbH & Co. The main focus of this project is the investigation, optimization and prediction of the biological degradation of TPH in the biopile process to enable the controlled operation of biopiles with heights of 4 to 5 meters. In this context, this static process is tested as a possible economic alternative to the usually-applied windrow process where the piles are turned periodically.

MATERIALS AND METHODS

To describe the biological processes in high biopiles, comparative investigations with aerated and unaerated biopiles were carried out. The biopiles were aerated with a pilot plant developed at the Department of Waste Management of the TUHH (figure 1). The system works as an oxygen-controlled high-pressure aeration system which enables aeration of the biopiles according to the respiratory activity which occurs in the biopiles due to microbial degradation processes. For this purpose, gas samples were taken automatically in regular intervals from different depths of the biopiles through sampling tubes and were analyzed for the oxygen concentration. The temperature at the continuous gas sampling sites was also recorded automatically. The oxygen values were used to generate oxygen profiles of the biopiles, and also to control the aeration. If the oxygen concentration drops below 2 Vol. %, the bio-pile is automatically aerated with a pulse of 1 m^3 of air with a pressure of 7.5 bar which flows into the soil within approximately 60 seconds. The concentration limit of 2 Vol. % oxygen for the beginning of the aeration is based on balancing investigations of the influence of the oxygen concentration on the biological degradation of TPH in soils and is slightly above the limiting concentration for the degradation process of 1 Vol. % (Hupe et al., 1998). In addition to the continuous oxygen measurements, in certain intervals gas samples were taken and analyzed by gas chromatography for O_2, CH_4, CO_2, volatile organic carbon (VOC) and trace gases. The TPH concentration, biomass, water content, and pH-value were determined in soil samples as well.

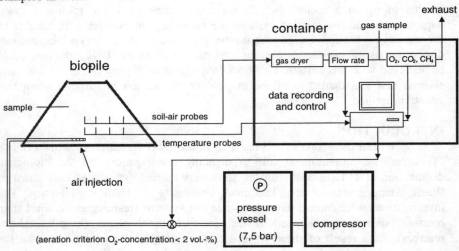

FIGURE 1. Pilot plant for the activity-controlled high-pressure aeration of biopiles.

To balance the TPH degradation process, carbon balances were simultaneously determined in lab-scale bioreactors. Those investigations were carried out in actively aerated solid state reactors with a volume of 6 litres (Koning et al., 1998). The different fractions of the initial total carbon content, TPH, biomass, organic carbon (C_{org}), carbon dioxide and VOC were determined and combined for the total carbon balance (Figure 7).

RESULTS

The investigations of the unaerated biopiles showed that the passive aeration of the biopiles was limited depending on the water content, the compaction ratio, and the biological activity of the soil. Figure 2 shows the penetration depth of oxygen for an unaerated bio-pile over the whole treatment period of 320 days.

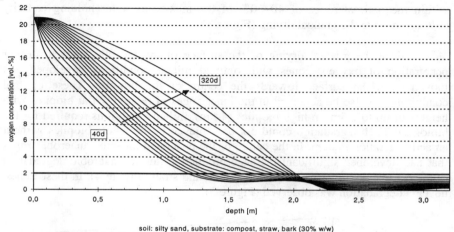

soil: silty sand, substrate: compost, straw, bark (30% w/w)
contamination: diesel fuel, temperature: 12 - 40°C, pH: 6.8 - 7.4, water content: 32 - 60% water-holding capicity

FIGURE 2. **Development of the oxygen concentration in the unaerated biopile (height: 3.2m).**

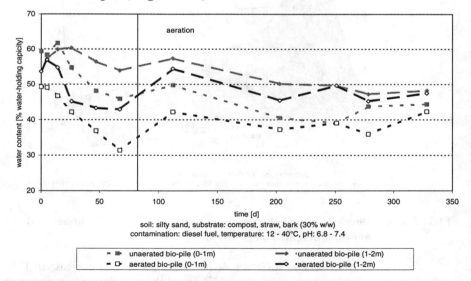

soil: silty sand, substrate: compost, straw, bark (30% w/w)
contamination: diesel fuel, temperature: 12 - 40°C, pH: 6.8 - 7.4

⬛ unaerated bio-pile (0-1m)	➡ unaerated bio-pile (1-2m)
◻ aerated bio-pile (0-1m)	◇ aerated bio-pile (1-2m)

FIGURE 3. **Development of the water content in the aerated and unaerated biopiles (height: 3.2m).**

As treatment time increased, the surface of the biopiles dried out and therefore the oxygen penetration depth increased. Despite this effect, oxygen concentrations in the optimal range for biological degradation were observed only in depths of less than 2 metres during the 320 days of the experiment. These observations agree with the estimations presented by Huesemann (1995) concerning the effects of passive aeration. The water content in the biopiles stayed within the optimal range of 40–60% of the maximum water holding capacity (WC_{max}) despite the gradual drying (Figure 3).

Since the gradual drying of the surface of the bio-pile leads to deeper penetration of oxygen, zones with optimal milieu conditions developed and moved from the surface towards the centre of the bio-pile. These optimal conditions can also be achieved in the inner zones of relatively flat biopiles with heights of up to 1.5m. In high biopiles (4-5m), milieu conditions required for the biological TPH degradation could not be achieved in the whole biopiles without an active aeration system. In the centre of the unaerated bio-pile, anaerobic zones that had developed significant methane production due to anaerobic degradation of the organic additives (30% compost, bark and straw; Figure 4). In these zones, no TPH degradation was observed (Figure 5).

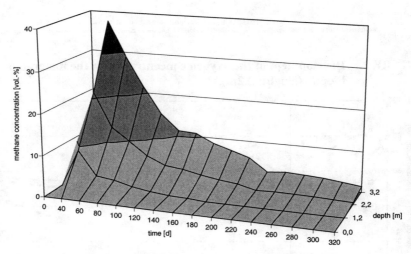

soil: silty sand, substrate: compost, straw, bark (30% w/w)
contamination: diesel fuel, temperature:12 - 40°C, pH: 6.8 - 7.4, water content: 32 - 60% water-holding capicity

FIGURE 4.   Development of the methane concentration in the unaerated bio-pile (height: 3.2m)

In the well-aerated surface zone (0-1m), a continuous decrease in TPH concentration could be observed after the 26th day of the experiment in the aerated as well as in the unaerated high biopiles. In contrast, a decrease in TPH concentration was observed in the centre (1-2m) of the aerated high bio-pile after the start of the aeration (84th day) and of the unaerated bio-pile after the 251st day of treatment. At the beginning of the experiment (84th day), the time interval between two aeration pulses was approximately 2 hours. With decreasing respiratory activity, the intervals extended up to 35 hours towards the end of the experiment (Figure 6).

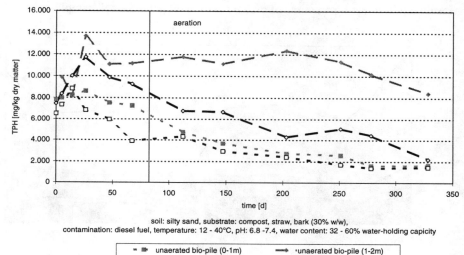

soil: silty sand, substrate: compost, straw, bark (30% w/w),
contamination: diesel fuel, temperature: 12 - 40°C, pH: 6.8 -7.4, water content: 32 - 60% water-holding capicity

▓ unaerated bio-pile (0-1m)	unaerated bio-pile (1-2m)
▢ aerated bio-pile (0-1m)	aerated bio-pile (1-2m)

FIGURE 5.   Development of the TPH concentration in the aerated and unaerated biopiles (height: 3.2m).

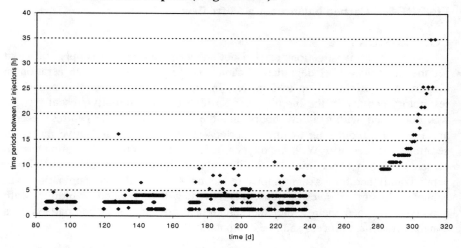

FIGURE 6.   Development of the intervals between aeration pulses.

The balancing investigations carried out simultaneously in lab-scale fixed-bed bioreactors show that less than 0.6% of the initial TOC was carried out as VOC in the gas phase as a consequence of the active aeration of the soil. This figure equals approximately 2.4% relative to the initial TPH contamination (Figure 7). The relatively high balance gaps are attributed to the inhomogeneous distribution of the organic additives (compost, bark, straw) in the soil, which then resulted in fluctuating amounts of organic carbon (C_{org}) in the soil samples.

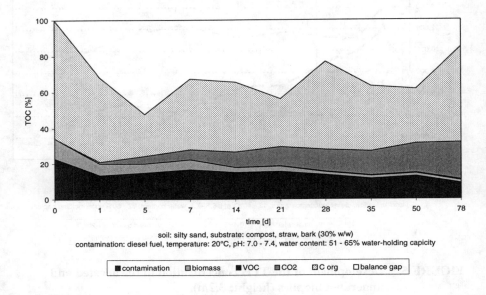

soil: silty sand, substrate: compost, straw, bark (30% w/w)
contamination: diesel fuel, temperature: 20°C, pH: 7.0 - 7.4, water content: 51 - 65% water-holding capicity

■ contamination   ▨ biomass   ■ VOC   ▤ CO2   ▨ C org   □ balance gap

FIGURE 7.   Carbon balance for a 6 litre fixed-bed reactor.

CONCLUSIONS

The investigations carried out show that an oxygen supply sufficient for the biological TPH degradation cannot be provided by passive aeration in biopiles with heights of up to 5 metres. In the case presented in this paper, the respiratory activity in the biopiles was so high that the initially-present oxygen was consumed within a few hours and the conditions in the centres became anaerobic. To avoid long treatment periods in technical plants, active means of aeration are necessary. With the oxygen-controlled high-pressure aeration system presented in this paper, biopiles with heights of 3 metres could be sufficiently aerated and therefore a continuous reduction of the TPH contamination was achieved. In further investigations, this system will be applied to biopiles with heights of up to 5 metres, and the economical advantages in comparison to a dynamic treatment process shall be demonstrated.

REFERENCES

Huesemann, M. 1995. Application of passive aeration to improve the cost effectiveness of bioremediation for petroleum contaminated soils.

Hupe, K., Heerenklage, J., Woyczechowski, H., Bollow, S., and Stegmann, R. 1998. Influence of oxygen on the degradation of diesel fuel in soil bioreactors. Acta Biotechnol. 18/98, S. 109-122.

Koning, M., Hupe, K., Lüth, J.-C., Cohrs, I., Quandt, C., and Stegmann, R. 1998. Comparative investigations into the biological degradation of contaminants in fixed-bed and slurry reactors. Proceedings of the 6th International FZK/TNO Conference, ConSoil 98', May 17-21, Edinburgh, U.K., p. 531-538.

BIOREMEDIATION OF DIESEL CONTAMINATED SOIL USING PILOT SCALE BIOPILES

Heleen D van Zyl (Technikon Pretoria, Pretoria, South Africa)
Leon Lorenzen (University of Stellenbosch, Stellenbosch, South Africa)

ABSTRACT: Three pilot scale biopiles and a landfarm to treat diesel contaminated sandy loam soil were constructed, operated and monitored for 113 days. The effects of vermiculite addition, increased bulking agent addition and tillage compared to forced aeration on contaminant reduction were examined. Samples were taken twice a week. Parameters monitored included total petroleum hydrocarbon (TPH) and microorganism concentrations, moisture content, temperature and pH. A degradation of 90 to 94 % diesel was achieved in the biopiles as well as the landfarm over 113 days. Target levels of 1000 mg TPH /kg soil were achieved within 30 days of treatment, indicating a reduction of 60%. Vermiculite, an increased amount of bulking agent and tillage had no significant enhancement on biodegradation rates achieved.

INTRODUCTION

Hydrocarbon contamination such as gasoline, diesel, oil and other petroleum products, has become a major environmental concern in South Africa. Aboveground spills at petrochemical complexes, overfilling of underground storage tanks and pipelines, as well as everyday operations at retail outlets have contributed to pollution of soil and groundwater. Although part of the leaked product can be recovered by drainage, pumping and extraction, a major portion often remains trapped in the pore spaces of the soil or bound to soil surfaces (Bhandari *et al.*, 1994). Gasoline components such as benzene, toluene, ethyl benzene and xylene isomers are especially hazardous wastes, and the presence thereof in the environment is of considerable public health and ecological concern due to their persistance, ability to be bioaccumulated, and toxicity to a wide variety of biological systems.

Petroleum hydrocarbons can be transformed into less-toxic compounds by means of a technology termed bioremediation. Microorganisms are used to clean up polluted soil by regulating the environment, to encourage microbiological activity towards enhanced degradation of contaminants. To facilitate the biotransformation process, variables such as temperature, pH, dissolved oxygen and nutrient concentration, and moisture content are adjusted and optimized for microbial metabolism.

Whenever soil contamination is great enough to be a potential source of groundwater pollution, soil will have to be removed and treated or disposed off. Biopiles offer a low-cost, low-maintenance *ex-situ* bioremediation treatment option for decontaminating large volumes of soil in a small area, under fully controlled conditions. Forced aeration is used to oxygenate the soil while the application of moisture and nutrients are regulated by means of overhead irrigation systems.

A 113 day pilot scale study examined the feasibility of using biopiling and landfarming to remediate diesel contaminated soil. Results from the initial contamination and the 113 day period of operation are presented.

Objective. The objective of this study was to determine the feasibility of using biopiling and landfarming to restore the quality of diesel contaminated soil to its original state, or at least, a state which is acceptable for future use. The effects of vermiculite addition, increased bulking agent addition and tillage compared to forced aeration on contaminant reduction were examined. The aim was to achieve target levels of 1000 mg TPH/kg soil. This indicates a reduction of 60% from the initial TPH concentration of 2500 mg/kg soil.

Site description. Approximately 5500 litres of diesel was spilled when a storage tank overflowed during the loading of bulk product at a petroleum depot. The majority of fuel was contained by a bund wall surrounding the fuel storage tanks. However, an unknown volume leaked from the bunded area and contaminated the neighbouring site. Diesel seeped into the top soil zone over a distance of approximately 30 m, to a depth varying form 0.5 to 1.2 m. The contaminated soil was excavated and transported to a biopile facility where it was left to weather over a period of three weeks and thereafter bioremediated using the biopiling and landfarming bioremediation techniques.

MATERIALS AND METHODS

A remote site was selected for construction of three biopiles and a landfarm to treat the diesel contaminated soil. A site of 10m x 10m was cleared of vegetation and prepared to have a slope of 2% to ease drainage of leachate. An enclosed 9m x 6m x 3m tunneltype greenhouse structure, divided into three sections of 6m x 3m x 3m (capacity : 20m^3 each), was constructed on the prepared site.

A leachate collection trench, sloped towards a collection sump was constructed across the width of each section. The leachate collection trench and sump, and the rest of the floor area of each division, were lined with a 1 mm thick high density polyethylene (HDPE) plastic liner to prevent groundwater contamination.

Slotted piping wrapped in geotech cloth, embedded in a 300mm thick layer of crushed stone underneath each biopile, acted as air distribution manifold. The aeration piping was connected to a rotameter and manual control valve to regulate the air flow rate, a biofilter to treat off-gasses and a 1 horsepower blower. The blower draws air through the pile to oxygenate it.

Each section was fitted with an overhead spray irrigation system which allowed for the controlled input of water. A cross-section of the biopile system is illustrated in Figure 1.

A landfarm (6m x 3m x 0.4m, capacity : 7.4m^3) was constructed adjacent to the biopile facility (Figure 2). The landfarm was lined with 1mm HDPE liner to prevent groundwater contamination. A rain gauge was put into place next to the landfarm to keep record of rain events.

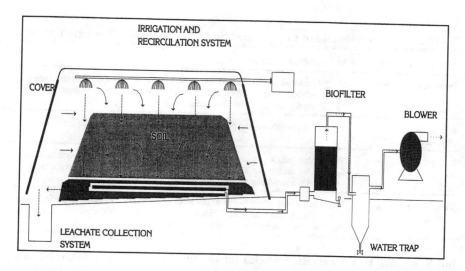

FIGURE 1.   Schematic Diagram of a Biopile.

FIGURE 2. Landfarm constructed adjacent to the Biopiles.

Soil preparation. To facilitate the biotransformation process, process variables such as the nutrient and moisture concentration are optimized for microbial metabolism. Materials added to the soil to optimize soil conditions for microbial activity are summarized in Table 1. The soil and amendments were measured and mixed thoroughly using a front end loader. The ratio of (soil + vermiculite/sand) : (sludge + bulking agent + MAP) was 5:1 on a volume basis for the landfarm, biopile B and biopile C, and 10:1 for biopile A. The mixture was transferred to the biopiles and landfarm and spread evenly. The soil was piled to a height of 1.2m within the biopiles and 0.4m within the landfarm.

TABLE 1. Materials added to Optimize Soil Conditions for Bioremediation.

MATERIAL ADDED	PURPOSE OF MATERIAL ADDED
Dried activated sludge	Sludge provides additional microorganisms, phosphorous and nitrogen to the soil. The dried sludge had a carbon content of 153 700 mg/kg, total nitrogen content of 29 750 mg/kg and total phosphate content of 83.1mg/kg.
Monoammonium phosphate (MAP)	MAP further increases the phosphate and total nitrogen content of the soil.
Bulking agent	Woodchips reduce compaction within the soil and maintain void space necessary for gaseous exchange. The bulking agent must not create a pile of unpractical volume nor overwhelm the volume of soil to be treated.
Vermiculite	Vermiculite was added to one of the biopiles (biopile B) to evaluate the influence thereof on moisture content. Vermiculite retains moisture and releases it slowly. Sand was used as inert material in the other two biopiles.

Soil Sampling and Analysis. Samples for chemical analysis were taken twice a week. A sample consisted of a composite of three samples taken across the heap, with each of these samples consisting of a mixture of soil taken at depths of 0.2m and 1m. For the landfarm, a sample consisted of a composite sample of two samples taken across the heap at depth of 0.02m. TPH concentrations were measured gravimetrically as well as by means of gas chromatography (GC) traces. Other parameters monitored included microorganism concentrations, moisture and oxygen content, pH and temperature.

RESULTS AND DISCUSSION

Soil TPH concentrations immediately after the spillage were 6000 mg/kg soil. After excavation, mixing and addition of bulking agent, sludge and MAP, TPH concentrations had declined to an average of 2200 mg/kg. The reduction can be attributed to inevitable dilution which occurred during excavation as well as the addition of amendments, biodegradation and volatilization occurring through weathering. However, the heavier petroleum fractions cannot be removed by weathering and had to be treated by means of biopiling.

Data show that soil contaminated with diesel derived petroleum hydrocarbons in the range of 2200-2500 mg TPH/kg soil could be bioremediated to below 300 mg TPH/kg soil in a time frame of 113 days (Figure 3). Target levels of 1000 mg TPH/kg soil were achieved within 30 days of treatment, indicating a reduction of 60%. The estimated half-life for the total petroleum hydrocarbon levels were 18 days for all biopiles and the landfarm.

Initially there was a very high degradation rate, however as the time increased, the degradation rate decreased and finally stabilized. This is due to the rapid degradation of alkanes initially. Heavier molecular weight compounds are more resistant to degradation resulting in a decrease in biodegradation rate as time increases. The initial biodegradation rate (days 0-13) for the landfarm was 160 mg TPHC/kg soil.day, which is slightly higher than for all biopiles, for which a biodegradation rate of 127-129 mg TPHC/kg soil.day was measured over this period.

This was due to the slightly higher moisture percentage within the landfarm due to heavy rainfalls. After 13 days of treatment the biodegradation rate declined to 13 ± 2 mg TPHC/kg soil.day for all biopiles and the landfarm and finally stabilized at a rate of 6 mg TPHC/kg soil.day due to the depletion of the carbon source towards the end of treatment.

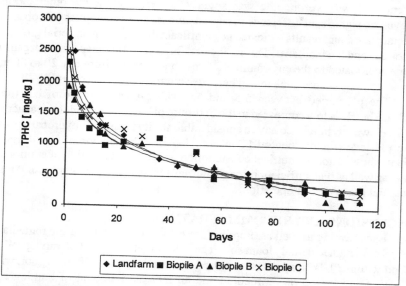

FIGURE 3. Total Petroleum Hydrocarbon Concentration vs Time.

GC traces further provided insight into what has occurred with the product. When biological activity was present, the envelope of n-alkanes began to decrease in size on the GC traces. Eventually all of the n-alkanes were removed. The remaining constituents appeared as a hump on the GC trace with a few discernable peaks (Zemo *et al.*, 1995).

A TPHC reduction of 94% for the landfarm, 91 % for biopile A, 94% for biopile B and 90% for biopile C was achieved. Other parameters measured during the period of operation are shown in Table 2.

TABLE 2. Parameters measured during operation

PARAMETER	BIOPILE A	BIOPILE B	BIOPILE C	LANDFARM
Moisture content (%w/w)	5-12%	5-17%	5-16%	3-23%
pH	6-6.6 (avg 6.2)	6-6.5(avg 6.2)	6-6.5(avg 6.2)	5.5-6.5 (avg 5.8)
Average nutrients ratio C:N:P (g/kg)	100:12:0.8	100:6:0.4	100:7:0.8	100:7:0.4
Oxygen concentration	6-10%	6-10%	6-10%	6-10%
Temperature (°C)	30-42	30-42	30-42	30-45

The moisture content was maintained within 30-80% of the maximum field moisture capacity (25%w/w) within all biopiles and the landfarm. The moisture content was less during the first 30 days of treatment due to extremely high daily temperatures (40°C). The watering program was altered to keep the moisture content within the region of 15% w/w. The moisture content of Biopile B which was supplemented with vermiculite was slightly higher than for the other biopiles. However, no significant difference in degradation rate for the biopiles was observed. Bacterial plate count results indicated no significant difference in bacterial growth for the biopiles and the landfarm. During the first 20 days of treatment microorganisms became acclimated to the environment. Growth proceeded from day 20 to 60 until the food source became depleted. Thereafter biomass slowly decreased.

The pH varied between 5.5 and 6.6 with a mean average of 6.2 for all biopiles. The pH is low compared to the optimum pH required for biological activity (6 to 8). However, fungi also have a considerable ability to degrade hydrocarbons and desire pH levels below neutral (Cookson, 1995). The analytical monitoring of nutrients showed normal nutrient consumption with no accumulation of ammonia or nitrate, as well as that sufficient mineral nutrients were available to microorganisms throughout the period of operation.

CONCLUSIONS AND RECOMMENDATIONS

Landfarming as well as biopiling are effective methods for the remediation of diesel contaminated sandy loam soil. The TPH target level of 1000 mg/kg soil was reached within 30 days, and was surpassed to reach less than 300 mg/kg soil within 113 days.The addition of vermiculite and an increased amount of bulking agent added, had no significant affect on biodegradation rates achieved. Similarly, a comparison between the uncontrolled landfarm and the controlled biopile system showed no significant increase in biodegradation rate. However, results for landfarming cannot be predictable and depends on environmental conditions such as high temperature (drying out soil) and rainfall.

In South Africa a need is recognised to build a centrally located regional bioremediation facility for petroleum contaminated soil, where soil can be brought in, microbes break down the contaminant and the soil is recycled. It is thus recommended that results from this study are used to aim future research towards this need.

REFERENCES

Bhandari, A., D.C.Dove and J.T. Novak. 1994. "Soil Washing and Biotreatment of petroleum Contaminated Soils". *Journal of Environmental Engineering*, 120(5) : 1151-1169.

Cookson, J.T. Jr, 1995. *Bioremediation Engineering : Design and Application,* McGrawHill.

Zemo, D.A., J.E.Bruya and T.E. Graf.1995."The Application of Petroluem Hydrocarbon Fingerprint Characterization in Site Investigation and Remediation", *Ground Water Monitoring and Remediation*, 15(2) : 147-156.

PERMANENT BIOPILE FACILITY AT TWENTYNINE PALMS MCAGCC

Keith A. Fields (Battelle, Columbus, Ohio
Thomas C. Zwick, Albert Pollack, and James Abbott (Battelle, Columbus, Ohio)
Chris Gonzalez (MCAGCC, Twentynine Palms, California)

ABSTRACT: Battelle has designed, constructed, and operated a permanent bioremediation facility (biopile) at the Marine Corps Air Ground Combat Center (MCAGCC) in Twentynine Palms, California. The facility consists of a 46-meter-long by 26-meter-wide by 18-centimeter-thick concrete pad underlain with a 60-mil high-density polyethylene (HDPE) liner. The concrete pad contains 12 troughs that are approximately 17 meters long, 18 centimeters wide, and 28 centimeters deep, which contain the perforated extraction pipe. Prior to operation, a soil treatability study was performed to determine optimal soil moisture content and nutrient levels. Three rounds of sampling (initial, interim, and final) were conducted on the first 1,900 m³ (2,500 yd³) batch of petroleum hydrocarbon-contaminated soil treated at the facility. During treatment of the first batch of soil, Battelle used an On-Line Environmental Monitoring System (OEMS) to monitor oxygen, carbon dioxide, total petroleum hydrocarbon (TPH), and pressure for soil gas and off-gas. Also, during operation of the biopile facility, a neutron probe was used to monitor moisture content inside the soil pile and soil pile temperature measurements were obtained. Insights into the design, construction, and operation of permanent biopile facilities have been attained.

INTRODUCTION

The Marine Corps Air Ground Combat Center (MCAGCC) is an active military facility, located in the Mojave Desert in south-central San Bernadino County, California about 8 kilometers north of downtown Twentynine Palms, which is about 87 kilometers north-northeast of Palm Springs. The site receives little rainfall and experiences a high evaporation rate. Near-surface soils consist of sands, gravels, silts, and clays originating from alluvial and playa deposits.

Soils were contaminated at MCAGCC during field training exercises, maintenance operations, and cleanup of areas around leaking fuel tanks. The petroleum-contaminated soil as currently known, is not considered hazardous waste or designated waste under Federal, state, or local laws. The primary sources of petroleum hydrocarbon-contaminated soil are as follows:

- JP-5 spills during storage and transfer operations around expeditionary air strip areas while using rubber bladders called tactical airfield fuel dispensing systems (TAFDS). About 70% of the affected soil was produced by spills at expeditionary air strips.
- Diesel fuel spillage occurring during fueling operations at the diesel fuel transfer station (DFTS) and from other fuel dispensing and

transfer equipment. About 20% of the affected soil was produced by diesel fuel spills.

- Petroleum-based lubricants and oils from vehicle and equipment maintenance operations and the investigation, cleanup, and removal of past petroleum releases. About 10% of the affected soil was produced by various lubricant and oil spills.

Soil from these sources has been excavated, transported, and stored at a designated contaminated soil staging area (CSSA) to await treatment.

METHODS

A permanent biopile facility was constructed within the confines of the MCAGCC Natural Resources Environmental Affairs (NREA) waste storage and treatment area during the spring of 1997. The permanent biopile facility was constructed to treat petroleum-contaminated soils in the CSSA as well as other contaminated soils.

Treated soil from the biopile facility is used as daily landfill cover at the Combat Center Landfill. The California Regional Water Quality Control Board designated that these soils would be acceptable for use at the landfill if total petroleum hydrocarbon (TPH) quantified as diesel (TPHd) concentrations were less than 1,000 mg/kg, benzene concentrations were less than 1 µg/kg, and ethylbenzene, toluene, xylene, and organic lead concentrations were less than 50 mg/kg.

Prior to treatment of the soils stockpiled at the CSSA, it was necessary to characterize the soil in the stockpile. To accomplish this task, 83 soil samples were collected from the CSSA during two sampling events (samples were collected for every 76 m^3 [100 yd^3] of soil in the stockpile). Results of the soil stockpile characterization indicated that the TPHd concentrations ranged from <1 to 15,000 mg/kg, benzene concentrations ranged from <0.001 to 0.1 mg/kg, and ethylbenzene, toluene, xylenes, and organic lead concentrations were less than the cleanup criteria of 50 mg/kg. Therefore, soil treatment effectiveness was driven by the ability to meet the TPHd and benzene cleanup criteria.

The biopile facility consists of a 46-meter-long by 26-meter-wide by 18-centimeter-thick concrete pad underlain with a 60-mil high-density polyethylene (HDPE) liner. The concrete pad contains 12 troughs that are approximately 17 meters long, 18 centimeters wide, and 28 centimeters deep, which contain the perforated extraction pipe. The extraction pipe has been placed in the troughs and covered with pea gravel so that heavy equipment can access the facility without damaging the piping.

Dump trucks and front-end loaders were used to form the soil pile approximately 1.8 meters from the back edge of the concrete curb wall on all four sides of the facility. The height across the width at the center of the pile was approximately 2.4 meters above the top of the slab. The total volume of soil placed on the biopile was approximately 1,900 m^3 (2,500 yd^3). To prevent compaction of the soil pile, the loader and dump trucks did not travel over soil after it had been placed on the pad (Figure 1).

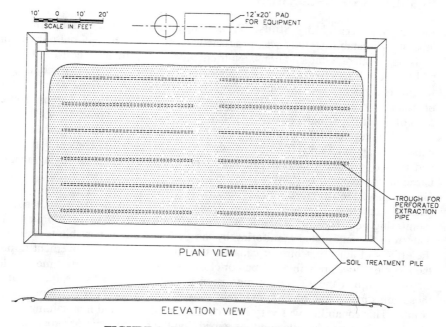

FIGURE 1. Permanent Biopile Facility

During formation of the soil treatment pile, water and fertilizer were added to the soil to obtain the optimal moisture and nutrient conditions determined by treatability study. Also, sample collection devices were installed to allow measurement of conditions within the pile. Sample collection devices included six aluminum access tubes for the neutron moisture probe, nine soil-gas sample collection screens and tubes, and thermocouples.

A dripline irrigation system was installed on the soil pile; however, it was not routinely used to augment soil moisture during the operation of the pile. It was operated only during irrigation testing activities. The dripline contains an orifice every 46 centimeters and each orifice releases water at a rate of 2.3 liters per hour if line pressure is maintained at 103 to 345 kPa. The irrigation lines were placed on top of the soil following formation of the pile.

Following formation of the soil pile and installation of monitoring and irrigation equipment, the biopile facility was operated and monitored for 6 months beginning September 10, 1997 and ending March 24, 1998. During the operation of the biopile, monitoring activities were performed to measure parameters that affected optimal bioremediation conditions in the soil and completion of the remediation process. The processes that were monitored included soil-gas, vacuums, blower exhaust, soil-moisture, and soil temperature. Also, in order to determine the treatment effectiveness, initial, interim, and final sampling of the soil pile was conducted.

RESULTS AND DISCUSSION

Laboratory Treatability Study. Prior to operation of the soil treatment pile, a laboratory treatability study was conducted to determine the optimal nutrient and moisture content concentrations. Nutrient addition primarily consisted of the addition of nitrogen (N) and phosphorus (P) using dry granular commercial fertilizers (urea and diammonium phosphate [DAP]). The optimal concentrations of nutrients represented a C:N:P (C represents the petroleum, or carbon, source) ratio of 100:15:1, which equated to 150 mg N/kg soil and 10 mg P/kg soil.

The results of the treatability study indicated that the optimal moisture content is approximately 10 % by weight. However, moisture content between 7 and 12 % by weight would provide sufficient moisture to enhance biodegradation based on the treatability study.

Operations Monitoring.   The Online Environmental Monitoring System (OEMS) monitored oxygen, carbon dioxide, TPH, and pressure at 9 soil-gas monitoring points installed within the soil pile and at the extraction piping. The data from the automated monitoring by the OEMS made it possible to determine the effects of adjusting blower capacities and the level of air movement through the biopile. It was possible to open or close the ball valves to permit more or less air flow through the 12 sets of extraction piping, and then to see the results of those adjustments in terms of the levels of oxygen in the soil gas at the monitoring points. This capability was very important for achieving and maintaining aerobic conditions in the entire pile and thus expediting the bioremediation process.

Respiration tests, useful in determining the rate of biodegradation, were used as an indication as to when the treatment was nearing completion. The trend observed for the two biopile respiration tests was that biodegradation rates decreased over time.

Blower exhaust was monitored by the OEMS and time-integrated, whole-air samples were collected for TO-3 and TO-14 analysis. TPH concentrations in the blower exhaust decreased from 350 ppmv at startup to 5.5 ppmv at shutdown. The collection of time-integrated, whole-air canister samples is recommended during biopile operations. They are easily collected and provide data to confirm the presence or absence of regulated emissions, while providing measurements of hydrocarbon discharge to the atmosphere from biopile operations.

Although soil moisture across the pile was in the range capable of supporting bioactivity, the variability in initial soil moisture levels (7 to 20% by weight), generally on the high end, can be attributed to limitations observed during manual hydration of the pile. During construction, the determination of how much water was being applied was not possible. After construction of the next soil pile, the moisture levels in the pile will be measured and the irrigation system will be used to obtain optimal moisture conditions.

When water was delivered to the surface of the biopile, pooling and runoff occurred. The slow infiltration rate associated with MCAGCC soil resulted in erosion of the sloped sides of the biopile and the delivery of a heavy soil load to the water containment system. In order to correct this situation, it will be necessary in future operations to enhance infiltration by setting the irrigation lines below grade. The irrigation system should be able to maintain elevated soil-moisture levels in the biopile, which will be critical during summer operations.

Soil temperatures ranging from 25°C to 35°C generally permit a maximum relative microbial reaction rate (Paul and Clark, 1989). If the soil temperature is <10°C or >55°C, then the rates can drop to below 25% of the maximum. Soil temperatures in the biopile were generally within a range that should not have impeded the biodegradation process. The logistics associated with either heating or cooling the biopile to maintain optimal microbial rates make soil temperature optimization impractical.

Soil Pile Sampling. Initial soil samples were collected using systematic grid sampling at random depths in the pile. The initial sampling grid was designed to give at least one sample for each 76 m^3 (100 yd^3) of material. During this initial sampling event BTEX and organic lead concentrations were below cleanup criteria and TPHd concentrations ranged from 930 mg/kg to 3,800 mg/kg. The 99% upper confidence level (UCL) of the mean concentration was 2,250 mg/kg.

For the final soil sampling event, samples were collected using systematic grid sampling at random depths in the pile. As specified by the RWQCB, the grid was designed to give at least one sample for each 38 m^3 (50 yd^3) of material. During the final sampling, BTEX concentrations were below cleanup criteria and TPHd concentrations ranged from 70.5 mg/kg to 5,800 mg/kg. The 99% upper confidence level (UCL) of the mean concentration was 1,130 mg/kg.

The mean and 99% UCL TPHd concentrations in the soil were significantly decreased during the 6 months of treatment. BTEX concentrations in soil were below the detection limits in all samples. The 99% UCL concentration for TPHd was approximately 13% above the cleanup criteria of 1,000 mg/kg. Therefore, it was recommended that the two 38 m^3 blocks of soil identified by the samples with the highest detected concentrations of TPHd (5,900 and 3,100 mg/kg) be combined with the next batch of soil treated at the biopile. Also, it was recommended that the soil pile, with the exception of the two 38 m^3 blocks with the highest TPHd concentrations, be considered suitable for use at the Combat Center Landfill.

CONCLUSIONS

Treatment of the first 1,900-m^3 batch of contaminated soil was successful. The calculated 99% UCL concentration for TPHd was reduced approximately 50% during the 6 months of operation. Also, BTEX concentrations in all final soil samples were below the cleanup criteria. Two soil samples collected during the final sampling event contained high TPHd concentrations that were shown not to represent the general population in the pile (i.e., to be outliers). Therefore, the two 38-m^3 blocks of soil represented by these samples will be retained for treatment during the next biopile operational period.

Optimization and operational conditions were identified during treatment of the first batch of soil at the biopile facility. Based on the knowledge gained during this project, the following recommendations are given for soils represented at the MCAGCC:

- The target nutrient concentrations determined from the treatability study result in a C:N:P ratio of 100:15:1. These target concentrations should be obtained by adding commercial fertilizers (DAP and urea).
- Based on the treatability study, soil moisture content should be maintained in the 7 to 12% range by using a drip-line irrigation system installed below the surface of the soil pile.
- Sufficient soil-gas monitoring can be accomplished using single-level oxygen sensors placed in the lower 25% of the soil pile. Nine, evenly spaced, oxygen sensors placed throughout the pile should be sufficient to provide statistically significant results. The comprehensive database provided by the OEMS during the operation of the first batch of soil was extremely useful; however, oxygen is the primary soil-gas field parameter required to adequately make decisions concerning the operation of the biopile. Vacuum measurements at the vapor extraction lines are recommended on a limited basis. The combination of oxygen and pressure monitoring will provide data to adjust flows from each vapor extraction line to maintain aerobic conditions throughout the pile. The oxygen sensors also will permit respiration testing.
- It is recommended that oxygen sensors not be installed directly above the vapor extraction lines or near the edge of the pile. The initial treatment pile monitoring indicated that these locations were subject to short-circuiting and provided limited useful data.
- Respiration testing should be used as a method for determining when final sampling should occur. When oxygen utilization rates drop to near background, final sampling should be initiated.
- Collection of time-integrated whole-air samples is recommended during subsequent biopile operations in order to provide off-gas loading estimates of regulated compounds.
- Soil temperature monitoring appears to be unnecessary due to temperatures being within the optimal range (>10°C and <55°C) and based on the impracticability of adjusting the temperature.
- Soil moisture monitoring should be conducted during the next biopile operational period that will occur during the summer months. Results from this monitoring will indicate water addition requirements during summertime operations. The first operational period did not indicate the need for irrigation.
- Both blowers should operate at full capacity unless off-gas discharge of petroleum constituents becomes a regulatory issue.

REFERENCES

Gilbert, R.O. 1987. *Statistical Methods for Environmental Pollution Monitoring.* Van Nostrand Reinhold, New York, NY.

Paul, E.A. and F.E. Clark. 1989. *Soil Microbiology and Biochemistry.* Published by Academic Press, Inc. San Diego, California.

DEVELOPMENT OF A NEW COVERING SYSTEM FOR THE ENVIRONMENTAL CONTROL OF BIOPILE IN COLD CLIMATE

Olivier Schoefs, Louise Deschênes and Réjean Samson
NSERC Industrial Chair in site bioremediation, École Polytechnique de Montréal
Montréal (Québec) Canada

ABSTRACT: A field study was undertaken with four 60 m^3 biopiles of contaminated soil to evaluate the influence of three types of covering system on the temperature inside the biopile. A semi-permeable black geotextile was compared to the two sealed double polyethylene membrane systems. Heat transfer was favored or restricted by choosing the color of the polyethylene membranes and by the presence of an insulating air layer between the two polyethylene membranes. Results showed that the air layer allowed to increase soil temperature up to a range that could enhance biodegradation. The white/white polyethylene membrane was considered to be the covering system offering the best performance since it allowed not only the temperature level to increase during fall conditions but also to reduce the temperature gradient within the biopile.

INTRODUCTION

Expanding industrial activities have led to petroleum hydrocarbon contamination of many sites located within extreme conditions restricting bioremediation from such places. In order to operate biopiles in spite of severe climatic conditions, the present study deals with the characterization of a new covering system that provides a mean for better temperature control.

Only a few studies in the literature have reported applications of above ground bioremediation in cold regions. As expected, in most cases, the average rate of biodegradation can be increased by increasing the soil temperature (Simpkin, 1995). Previous studies have also shown that bioremediation was possible in cold climates (Moore, 1995; Sayles, 1995). Sayles et al. (1995) presented three soil warming methods: 1) the application of warm water at a low rate, 2) the enhancement of solar warming by using clear plastic in summer and insulation in winter (solar/insulation warming method), and 3) the use of buried heat tape. The efficiency of the solar/insulation warming method was low compared to the other methods but the technique proved to be very interesting as the costs of installation and operation were very low. In view of this important economic advantage, the concept of solar warming was retained for this study and applied to biopile treatments.

MATERIALS AND METHODS

Biopile design. Four 60 m^3 soil biopiles (9 m x 4.5 m x 1.5 m) were built using a soil contaminated by semi-volatile petroleum hydrocarbons (diesel). The soil characterization showed a level of contamination of 5,400 mg/kg of mineral oil and grease. All biopiles were equipped with an aeration system and with thermocouples and Time-Domain-Reflectometry (TDR) probes placed at three different depths (40, 80 and 120 cm from the top of the biopile) for temperature and water content monitoring respectively. A data acquisition system collected data which were recorded every two hours. Details are available in Schoefs et al. (1998).

Covering System. The new biopile covering system consisted of a polyethylene double membrane. Each membrane was 0.2 mm thick. Depending on the biopile, the membrane used was white, black or translucid. The first biopile was equipped with the conventional covering system which is a semi-permeable black geotextile membrane. The second one was covered first with a black polyethylene membrane, followed by the upper layer which consisted of a translucid membrane to enhance heat transfer. In order to restrict the heat transfer by solar radiation, the last two biopiles were covered each with two white polyethylene membranes. A 20 cm thick insulating air layer was maintained between the two polyethylene membranes with an air pump (capacity of 70 m^3/h) which also introduced air between the soil and the first polyethylene membrane (Figure 1). A sealing system, named Polylock™, maintained the air-seal around the biopile.

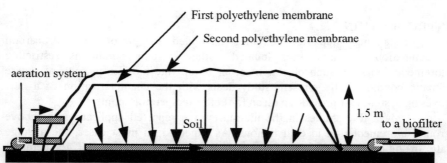

FIGURE 1. Schematic representation of a biopile equipped with a double polyethylene covering system

Some of the innovative features of the polyethylene double membrane lie in the thickness of the air layer and the choice of the membrane color. Compared with the commonly used black geotextile membrane (Figure 2a), the use of a white polyethylene double membrane allows the air layer to play the role of an insulating blanket (Figure 2b). In contrast, when a black/translucid polyethylene double membrane was used , the black surface was heated up by solar radiation which transmitted heat to the air layer, and thus played the role of a heat sink (Figure 2c).

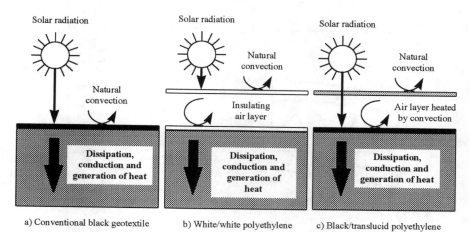

a) Conventional black geotextile   b) White/white polyethylene   c) Black/translucid polyethylene

FIGURE 2. Heat transfer through covering systems and soil

Initial soil characterization of the biopiles. The granulometry analysis showed that the soil was a sandy loam containing 5% clay, 23% silt and 72% sand. A bulk density of 1650 kg/m^3 proved that compaction of the soil was low, resulting in enhanced water and air permeability. A water holding capacity of 0.21 g of water/g of dry soil indicated a good ability of the soil to retain water. The soil pH (7.65) was found to be within the biodegradation optimum reported in the literature, which is 7.5 to 7.8 (Dibble, 1979). In contrast, the C/N ratio of 40 was shown to be higher than the optimum value of 10 reported in the literature (MEF, 1995), which indicates that nitrogen could be a limiting factor for biodegradation. The soil water content increased with depth in the biopile. Results showed a value of 15% v./v. at 40 cm, 20% v./v. at 80 cm and 30% v./v. at 120 cm from the top of the biopile. This humidity gradient can be explained by the gravity force exerted on water and by the air flow of the aeration system. The suggested water content is found to be within the range of 25% to 85% of the water holding capacity (MEF, 1995), that is, for the studied soil, a range of water content between 10% v./v. and 32% v./v.. The biopile water content was thus favorable but not necessarily optimal for biodegradation. Details are available in Schoefs et al. (1998).

RESULTS AND DISCUSSION

Effects of covering system on soil temperature. The type of the covering system used (black geotextile, black/translucid polyethylene membranes, and white/white polyethylene membranes) has an important effect on the average biopile temperature and daily fluctuations (Figure 3). This effect was more or less pronounced depending on the selected time period (summer and fall).

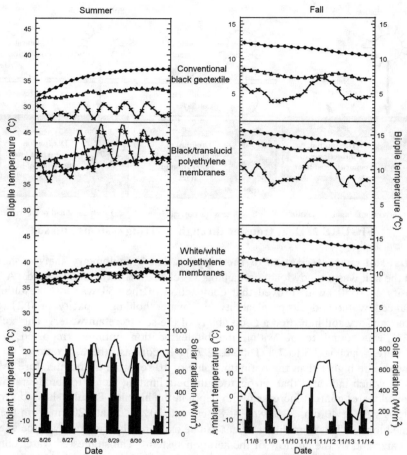

FIGURE 3. Climatic conditions and temperatures in biopiles
(◊ 120 cm, Δ 80 cm, ✕ 40 cm)

Compared to the conventional covering system (single black geotextile membrane) and for the two different periods (summer and fall), the average soil temperature increased when the new double polyethylene membrane system was used. Indeed, while the single black geotextile membrane provided the lowest average temperature in the summer (about 32°C) as well as in the fall (about 8°C), the use of the black/translucid polyethylene membranes allowed the average temperature to increase to about 42°C in the summer and 13°C in the fall. These observations mean that the air layer not only contributed to insulating but also to heating the soil. The white/white polyethylene membranes provided intermediate average temperature which was about 37°C in the summer and 12°C in the fall. Compared with the conventional black geotextile, the increase in the average soil temperature was about 5°C for both studied periods.

This result allows heating and insulating properties of the double polyethylene membranes to be segregated, depending on the membrane color. Indeed, the white/white polyethylene membranes can be considered as an insulating thermal resistance and the black/translucid polyethylene membranes as an insulating and heating thermal resistance.

The temperature profile inside the biopile was also influenced by the type of covering system. For example, while, during both summer and fall, a difference of 10°C was observed between 40 cm and 120 cm depths inside the conventional biopile and the one equipped with the black/translucid polyethylene membranes, the temperature difference observed using the white/white polyethylene membrane was reduced to 5°C in the summer and 8°C in the fall. This observation can be explained by considering the difference between heat transfer characteristics of the two covering systems. As mentionned before, the black/translucid polyethylene membranes absorbed more thermal energy than the white/white polyethylene membranes leading to more thermal energy being transmitted to the soil. Therefore, since the heat capacity of the soil is large, heat transfer inside the soil was very low compared to the variability of the heat flux transmitted by the covering system. This means that, within the biopile equipped with the white/white polyethylene membranes, the soil receiving a smaller heat flux can more easily reach a steady state temperature which is shown by the small temperature gradient as a function of depth. In the same way, the biopile equipped with the black/translucid polyethylene membranes received a relatively large heat flux and the thermal system could not reach a steady state temperature. This is confirmed by the important temperature gradient observed as a function of depth.

The last observation concerns the effect of the different covering systems on the daily temperature fluctuations. This effect can only be observed at the top of the biopile since, in all biopiles, daily temperature fluctuations were negligible below 80 cm from the top of the biopile. Therefore, at 40 cm depth, the largest fluctuations were observed for the biopile equipped with the black/translucid polyethylene membranes ($\Delta T \approx 10°C$ in the summer and $\Delta T \approx 2°C$ in the fall). Daily temperature fluctuations in the biopile equipped with the black geotextile were smaller ($\Delta T \approx 3°C$ in the summer and $\Delta T \approx 0.5°C$ in the fall) and the ones in the soil of the biopile equipped with the white/white polyethylene membranes were even lower ($\Delta T \approx 2°C$ in the summer and $\Delta T \approx 0°C$ in the fall). These observations have led to two conclusions. First, the color of the covering system influences temperature fluctuations at the top of the biopile especially in the summer when solar radiation was more intense. In particular, the black/translucid polyethylene membrane presented higher temperature fluctuations than the white/white one. Second, the presence of the air layer resulted in an increase in the temperature fluctuations. This can be explained by the fact that the air layer absorbed a large amount of solar energy leading to an important temperature increase very quickly during the day. Then in the evening, as the temperature inside the biopile is high, the decrease is also important.

In conclusion, the double polyethylene membrane succeeded in increasing the temperature inside the biopile during both summer and fall. This means that, as expected, biodegradation could be enhanced in spite of cold climatic conditions. It is also interesting to note that, in summer, the temperature fluctuated around 40°C in the biopile equipped with the black/translucid polyethylene membranes, which is probably a upper limit for the biodegradation temperature.

However, despite the difference in temperature within the soil piles based on cover type, the degree of biodegradation was essentially the same for all biopiles. This observation is attributed to the range of contamination faced during the experiment (from 5400 down to 2500 mg/kg) in which biodegradation was not limited by microbial and activity factors such as temperature but by contaminant desorption and diffusion considerations (De Jonge, 1997).

ACKNOWLEDGEMENTS

The authors acknowledge the financial support from the industrial Chair partners: Alcan, Bodycote/Analex, Bell Canada, Browning-Ferris Industries, Cambior, Centre Québécois de valorisation de la biomasse et des biotechnologies (CQVB), Hydro-Québec, Natural Science and Engineering Research Council (NSERC), Petro-Canada and SNC-Lavalin.

REFERENCES

De Jonge, H., Freijer, J. I., Verstraten, J. M. and Westerveld, J. 1997. "Relation between Bioavailability and Fuel Oil Hydrocarbon Composition in Contaminated Soils". *Environ. Sci. Technol.*31, 771-775.

Dibble, J. T. and Bartha, R. 1979. "Effect of Environmental Parameters on the Biodegradation of Oil Sludge". *Appl. Environ. Microbiol.* 37 (4), 729-739

MEF 1995. Lignes directrices pour le traitement de sols par biodégradation, bioventilation ou volatilisation. Sainte-Foy, Québec

Moore, B. J., Armstrong, J. E., Barker, J. and Hardisty, P. E. 1995. "Effects of Flowrate and Temperature During Bioventing in Cold Climates". *Third International In Situ and On-Site Bioreclamation Symposium* (San Diego, California). Batelle Press. 3(2). 307-313

Sayles, G. D., Leeson, A., Hinchee, R. E., Vogel, C. M., Brenner, R. C. and Miller, R. N. 1995. "Cold Climate Bioventing with Soil Warming in Alaska". *Third International In Situ and On-Site Bioreclamation Symposium* (San Diego, California). Battelle Press. 3(2). 297-306

Schoefs, O., Deschênes, L. and Samson, R. 1998. "Effeciency of a New Covering System for the Environmental Control of Biopiles Used for the Treatment of Contaminated Soils". *Journal of Soil Contamination.* 7(6), 753-771.

Simpkin, T. J., Carothers, G., Hoffman, R. W., Eldler, R. and Collver, B. F. 1995. "The Influence of Temperature on Bioventing". *Third International In-Situ and On-Site Bioreclamation Symposium* (San Diego, California). Battelle Press. 3(2). 315-322.

METABOLIC PRODUCTS FROM PETROLEUM HYDROCARBONS IN OIL POLLUTED SOIL

Augusto Porta, Antonino Trovato (Battelle, Geneva, Switzerland)
Jerry M. Neff (Battelle, Duxbury, Massachusetts)
Giorgio Andreotti (ENI-Agip, Milano, Italy)

INTRODUCTION

Following the Trecate No. 24 oil well blowout, Battelle performed a biodegradation study of total petroleum hydrocarbons (THPC) and polycyclic aromatic hydrocarbons (PAH) in heavily oiled soils. Studied soils included those located in agricultural fields and those that had been removed from the fields and subjected to remediation in two biopiles (designated biopiles 1 and L in the Figures) with a combined volume of 30,000 m^3.

The crude oil released from the Trecate well is a medium weight oil with a relatively high flash point and wax content. The concentration of total resolved and unresolved petroleum hydrocarbons in fresh Trecate 24 oil is approximately 830 mg/kg (83%). The remaining 17% of the mass probably includes compounds with a boiling point above that of n-C$_{40}$-alkane or below n-C$_8$-alkane. Total resolved PAHs (naphtalene through benzo(g,h,i)perylene) represent just over 1% of the oil hydrocarbon content. Most of the unresolved hydrocarbons are a complex mixture of branched and cyclic alkanes and naphtheno-aromatic hydrocarbons that constitute the unresolved complex mixture of most crude petroleum products.

PRIOR BIOPILE TREATMENT

Prior studies of the two biopiles show that substantial microbial degradation occurs in linear and cyclic alyphatic hydrocarbon compounds. Figure 1 summarizes the evolution of average TPHC concentrations in the two biopiles over time. Figure 2 depicts the evolution of the average concentration of the total resolved hydrocarbons in these biopile soils.

Recently, few studies have been carried out that elucidate the decomposition pathways of these compounds and show that organic acids and ketones (alicyclic and branched-chain aliphatic organic acids and diacids) are formed.

The rate and extent of degradation of PAHs in biopile soils varied in relation to the class of PAHs (i.e., from 2 to 6 ring PAHs), as can be seen in Figures 3 and 4. Microbial metabolic pathways for degradation of PAHs containing up to three rings have been proposed, while less information has been available about metabolism of the larger, more recalcitrant, high molecular weight PAHs.

Most of the information available in the literature on metabolites resulting from PAH microbial degradation derives from studies using culture media in microcosm bioreactors: little or nothing has been done to identify metabolites on polluted, weathered soil samples containing complex mixtures of PAHs and PAH metabolites [3].

Because of the extreme complexity of the non polar and slightly polar organic fractions in an agricultural soil contaminated with crude oil, it is very difficult to identify chemicals in soil extracts that clearly appear as degradation products by microbial metabolism and photooxidation of these hydrocarbons.

MATERIALS AND METHODS

First, oil-contaminated soils in agricultural fields were sampled in March 1994, about one week following the oil well blot out, at two monitoring stations located close to the oil well (Stations 47 and 15) in order to identify broad classes of compounds. There after, biopile soils were sampled more extensively. A description of analyses performed on both types of soils follows.

The presence of major classes of petroleum metabolites and the concentrations of selected individual metabolites were measured in soil samples using GC/MS techniques. Metabolites were isolated from soils using a solvent extraction and a step-wise fractionation procedure. Homogenized soil was acidified with HCl, desiccated with anhydrous sodium sulfate, fortified with surrogate internal standards, and then serially extracted with dichloromethane using Tecator Soxhlet methods. Metabolites were separated from the extract using a mixed-bed 2% deactivated alumina-2% deactivated silica fractionation column. Five discrete fractions containing various classes of hydrocarbons and metabolites were isolated from the cleanup column: F1-100% heptane (saturated hydrocarbons); F2-8:7 DCM:heptane (PAH); F3-100% DCM (diketones, carboxaldehydes and PAH metabolites); F4-100% methanol (alcohols); F5-5% sulfuric acid (organic acids). Each fraction was spiked with a unique deuterated recovery internal standard and analyzed by full scan GC/MS. Metabolites in each class were identified by retention time and spectral matching relative to known standard compounds. Response factors for most metabolites were assigned relative to a surrogate compound of similar chemical structure found in the five-point calibration solution.

RESULTS

Field Soils. Five classes of possible degradation products were detected in soil samples from the two stations (Table 1). These were alcohols, organic acids, alkenes, sterols/sterones, and alkanes.

At the time of the field soil sampling the most abundant degradation products at Station 47 were organic acids. The most abundant degradation products in soils from Station 51 were sterols and sterones. These alcohols and organic acids typify initial degradation products associated with some of the alkanes in the crude oil.

Identification of organic chemicals detected in soil extracts was performed by comparing the mass spectra of the individual analytes with the mass spectra from a reference library. A better than 85% match between the analyte mass spectrum and a library mass spectrum was considered a likely identification; however, some compounds may have been misidentified.

The sterols and sterones are not expected degradation products of crude oil hydrocarbons. However, some of them resemble some sterane and triterpane components of crude oil and could represent their degradation products. Alternatively, these sterols and sterones could be either microbial biomass components, indicative of microbial activity in the soils and thus possibly related to hydrocarbon degradation, or they could be natural soil biochemicals. Although many bacteria do not contain significant concentrations of sterols, fungi do, and both types of soil and water microbes are capable of degrading several types of sterols [1]. Soils and sediments usually contain complex assemblages of sterols; sterones are oxidation products of sterols [2]. Most of the sterols in soils come from decaying plant material. Several sterols are abundant in rice bran, including cholesterol, stigmasterol, and sitosterol. Stress from chemical contaminants may cause alterations in sterol concentrations and ratios in rice shoots. Thus, the sterols in rice field soils near Trecate could be natural soil components or direct or indirect degradation products of petroleum hydrocarbons.

TABLE 1. Relavite concentration ranges of organic chemicals (possible hydrocarbon degradation products) detected in organic extracts of soils from Stations 47 and 51 during the preliminary field survey in March 1994. Concentrations are expressed as the ratio of each analyte to the internal standard androstane.

Chemical	Station 47	Station 51
Heptanol	0.059	0.108
1-Eicosanol	0.134	0.161
Total alcohols	0.194	0.269
Tetradecanoic acid, methyl ester	0.093	0.004
Hexadecanoic acid, methyl ester	0.363	0.048
9,12-Octadecanienoic acid, methyl ester	0.573	0.060
9-Octadecanoic acid, methyl ester	0.705	0.030
Octadecanoic acid, methyl ester	0.035	0.011
Total organic acids	1.769	0.153
1-Docosene	0.104	0.111
Total alkenes	0.104	0.111
Cholesterol	0.014	0.241
Cholestanol	0.007	0.050
Coprostan-3-one	0.015	0.037
Ergost-5-en-3-ol	0.031	0.334
Stigmasterol	0.054	0.543
Gamma-Sitosterol	0.084	0.445
Ergostanol*	0.019	0.066
24S-Ethylcholesta-4,22E-dien-6-one*	0.039	0.117
Sigmast-4-en-3-one	0.045	0.136
Total sterols-ones	0.309	1.971
Eicosane	0.024	0.050
Cyclooctacosane	0.059	0.214
Total alkanes	0.083	0.263
Total degradation products	2.549	2.767

* less than 85% match with mass spectra library.

Some authors reported detecting large numbers of cyclic, aromatic, and aliphatic acids produced during microbial degradation of diesel fuel and lubricating oil in artificial soils. Others reported the presence of a large number of low molecular weight organic acids in ground water down-gradient from a crude oil spill. The aliphatic acids reported by these authors were similar to the organic acids detected in soils from Stations 47 and 51. No cyclic acids (e.g., cyclohexane carboxylic acid) or aromatic acids (e.g., naphthalene-1-carboxylic acid) were detected in Trecate soils. It is probable that these acids are relatively unstable in the soil and are either washed away rapidly or degraded further by co-metabolic reactions initiated by the complex assemblages of hydrocarbon-degrading bacteria and fungi in soil.

The alcohols, fatty acids, and sterols, detected in Trecate soils may have been derived in part from the rice crops grown there and from microbial activity in the soils. However, it is probable that some of these organic chemicals also were derived directly or indirectly from the biodegradation and photooxidation of petroleum hydrocarbons from the spilled oil.

Metabolites in Biopile Soils. Three major classes of petroleum metabolites consisting of ketones, aldehydes, and organic acids were detected in degraded soils from the two biopiles. Organic acids represent by far the largest classes of metabolic products found in these soils (Figure 5). Of these, two diagnostic metabolites, 2-tridecanone and tetracosanoic acid, were chosen as target indicators for metabolite production during petroleum degradation. Unweathered Trecate oil, background soil, and wood chips added to the biopiles contained none of the target metabolite compounds, demonstrating that any metabolites found in the biopile soils were produced during petroleum degradation.

In periodic analyses of biopile soils, the two diagnostic metabolites show relatively high starting concentrations (biopiles soils had already undergone substantial degradation in the field prior to before biopile treatment) which tend to decrease with time (Figures 6 and 7).

CONCLUSIONS

This work shows that various metabolites are formed during oil degradation in agricultural soils, the largest portion of which consists of organic acids. Because relatively little is known about the toxicity of such hydrocarbon metabolites and because these may be more toxic than the parent compound, the analytical results from this study should be complemented with various ecotoxicity studies in order to more completely evaluate the impact of the soil bioremediatiion. This may be particularly important in deciding the future use of the biopile soils.

REFERENCES

[1]   Langbehn, A., and H. Steinhart H. 1994. "Determination of organic acids and ketones in contaminated soils". *Jour. High Resol. Chrom.,* Vol. 17.

[2]   Weismann, W.H. 1995. "Total petroleum hydrocarbon criteria working group: A risk-based approach for the management of total petroleum hydrocarbons in soil". *Jour. Soil Contam.,* 7(1):1-15.

[3]   Zink, G., and Lorber, K.E. 1995. "Mass spectral identification of metabolites formed by microbial degradation of polycyclic aromatic hydrocarbons (PAH)". *Chemosphere,* Vol. 31, No. 9, pp. 4077-4084.

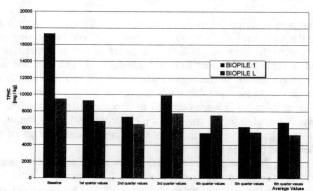

FIGURE 1. Average TPHC concentrations for biopiles 1 and L.

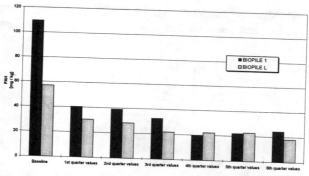

FIGURE 2. Average DTPH concentrations for biopiles 1 and L.

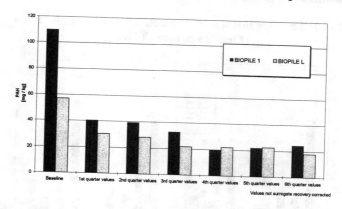

FIGURE 3. Average PAH concentrations for biopiles 1 and L.

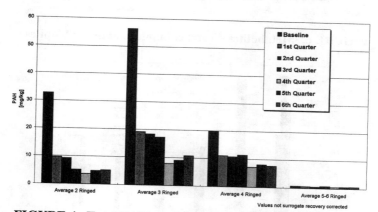

FIGURE 4. Total PAH degradation by ring size for Biopile 1.

	Reference Oil	bp1z3nl1		bp1z3nl3		bp1z3nl5		bplz3el2		bplz3el3		bplz6wf2		bark
		bp01	bp11	bp02	bp12	bp03	bp13	bl01	bl11	bl02	bl12	bl09	bl19	
Benzoic Acid												x	x	x
Decanoic Acid														x
Dodecanoic Acid													x	x
Tridecanoic Acid														x
Tetradecanoic Acid					x	x	x	x		x			x	x
Pentadecanoic Acid				x	x	x	x	x		x			x	x
Hexadecanoic Acid		x		x	x	x	x	x	x	x	x	x	x	x
Heptadecanoic Acid				x	x	x				x	x	x	x	x
Octadecanoic Acid		x	x	x	x		x	x	x	x	x	x	x	x
Nonadecanoic Acid														x
Eicosanoic Acid		x	x	x	x	x	x	x		x	x		x	x
Henicosanoic Acid			x				x							x
Docosanoic Acid		x	x	x	x	x	x	x		x	x	x	x	x
Tricosanoic Acid		x	x	x	x	x	x	x		x		x	x	x
Tetracosanoic Acid		x	x	x	x	x	x	x	x	x	x	x	x	x
Pentacosanoic Acid			x			x		x		x			x	x
Hexacosanoic Acid			x	x	x	x		x		x			x	x
Heptacosanoic Acid														
Octacosanoic Acid			x		x	x	x	x		x	x		x	x

Presence in sample indicated with a "x"

FIGURE 5. Organic acids in selected biopile samples from Oct. '95 to March '96

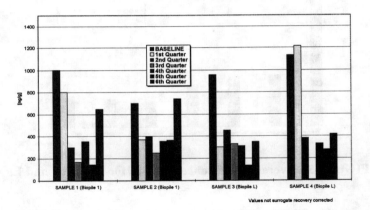

FIGURE 6. Metabolites - Tetracosanoic acid in the biopiles

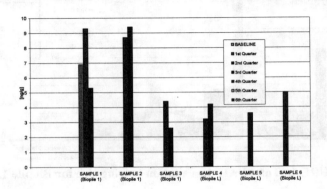

FIGURE 7. Metabolites - 2-Tridecanone in the biopiles

HYDROCARBONS BIODEGRADATION IN LANDFARMING SOILS CONTAINED IN CONTAINERS

Claudio O. Belloso (Facultad Católica de Química e Ingeniería, Rosario, Argentina)

ABSTRACT: A pilot-scale test was performed, consisting on landfarming soils inoculation with different isolated bacteria strains of the same contaminated soil with hydrocarbons. The goal of the project was to determine which bacteria type and under what conditions the biggest degradation of hydrocarbons is gotten. Different nutrients concentrations were applied. All soils were maintained outdoors for 90 days. The biggest biodegradation rate obtained was 45% in one of the soils tested.

INTRODUCTION

The hydrocarbons biodegradation in soils is an alternative way that can be used both for the treatment or final disposition of wastes produced by petroleum refineries.

It is now accepted that the capacity of autodepuration of soils is limited and that the activities of its microbial population, true motor of the biodegradation of wastes, is repressed with the massive and irrational incorporation of wastes, either for the high concentrations that retard its degradation or for its condition of dangerous wastes.

The well-known technologies as landfarming, land treatment or land application, are methods for remediation of petroleum hydrocarbons through biodegradation. These technologies consist on the application the hydrocarbons-contaminated soil or in the application of sludge of tank bottom of raw petroleum in a fine layer on another prepared soil for such an end.

Hydrocarbons concentrations more than 10% are extremely inhibitory of the biodegradation process (Leahy and Colwell, 1990).

Incorporation of hydrocarbons to the soil alters the relationships carbon/nitrogen or carbon/phosphorus to values that they can be unfavorable for microbial growth. The nitrogen and phosphorus availability can be limiting factors (Atlas and Bartha, 1992).

The effectiveness of these methodologies depends on countless factors, among them, the agronomic, topographical and microbial characteristics of the receiving soil, characteristics and composition of the applied wastes, climatic conditions, etc.

Objective. To compare the biodegradation of different bacterial strains in landfarming soils contained in containers.

MATERIALS AND METHODS

Soil was obtained out of landfarming belonging to Petroleum Refinery San Lorenzo S. A. located in San Lorenzo City, Santa Fe Province, Argentina.

Initial hydrocarbon in this soil was 5% w/w (expressed as ethylic ether extract).

TABLE 1. Percentual composition of hydrocarbon residual contained in landfarming soil.

Hydrocarbon distribution	Percentage
Saturates	44
Aromatics	17
Resins	14
Asphaltenes	25

Initial Total Organic Carbon of the landfarming soil was 1.07% w/w. Soil was characterized as sandy loam (texture: 61% sand, 16% silt, 23% clay).

Eleven metallic containers were built in which 30 Kg of soil of landfarming per container were put. To avoid contact of the soil with the metal of the container, proceeded to recover the containers inner walls with a thick layer of non-biodegradable resin under the conditions of the test.

Three bacterial strains for the inoculations were chosen, previously cultured in a liquid mineral medium with ethereal extract (EE) as sole carbon source.

The EE was obtained by extraction landfarming soil during 6 hs. with ethylic ether using a Soxhlet apparatus.

TABLE 2. Liquid Mineral Medium (LMM) composition.

NaCl	5.0 g
$MgSO_4.7H_2O$	0.2 g
KNO_3	3.0 g
K_2HPO_4 anhydrous	1.0 g
$(NH_4)_2 SO_4$	1.0 g
$(NH_4)H_2PO_4$	1.0 g
Distilled water.	1000 ml

The cultures were performed in reactors made of cylindrical recipients of polypropylene of 1.5 lt. of capacity with a continuous agitation system.

The biomass free upper liquid was weekly discarded and the liquid mineral medium with the carbon source was reloaded.

Bacterial strains named W1, W2 and W3 of the same soil of the landfarming were isolated, since we were unable to identify them exactly. (Belloso et al., 1997).

Just one application of fertilizers NPK type at the beginning of the trial were made.

Fertilizers according to the proportions shown in the table 3 were formulated and applied.

TABLE 3. Fertilizers application and formulations.

Denomination	Relationship C/N	Relationship NPK	Source of nitrogen
Z1	20	20:2:1	NH_4^+
Z2	100	20:20:1	NH_4^+

Containers inoculation. Soil, fertilizer and bacterial strain in each one of the containers were put according to that shown in table 4.

TABLE 4. Soil quantity, fertilizer and bacterial strain in each one of the containers.

Container N°	Soil (Kg)	Hydrocarbons initial concentration (% w/w)	Abiotic control	Inoculated strain	Fertilizer
1	30	5.0	-	-	-
2	30	5.0	$HgCl_2$	-	-
3	30	5.0	-	W1	-
4	30	5.0	-	W2	-
5	30	5.0	-	W3	-
6	30	5.0	-	W1	Z1
7	30	5.0	-	W1	Z2
8	30	5.0	-	W2	Z1
9	30	5.0	-	W2	Z2
10	30	5.0	-	W3	Z1
11	30	5.0	-	W3	Z2

Container N°1 contains only soil as control test, without fertilizer and without bacteria. Container N°2 contains soil and $HgCl_2$ (2g/100g of dry soil) in order to evaluate abiotics losses of hydrocarbons (Ferrari et al., 1994).

Containers were maintained outdoors for 90 days since July 1996 to October 1996 once inoculated.

The average temperature was 18°C. Soils were aired weekly by blend the whole content.

At the beginning of the trial and every 30 days the following parameters in each soil were determined: pH, moisture, total hydrocarbons, total aerobic heterotrophic bacteria (TAHB) and hydrocarbons degrading microorganisms (HDM).

The pH in suspensions of soil (40% w/v) in $CaCl_2$ 0.01M solution was measured.

Moisture in soils samples was determined by drying to 105°C until constant weight.

The hydrocarbons content was determined by extraction in Soxhlet with Ethylic Ether for 6 hs.

TAHB in soils were enumerated by the pour plate method (Clesceri et al., 1992) using Merck plate count agar. All plates were incubated at 30°C for 24 hours.

HDM in soils were enumerated by the more probable number method (Clesceri et al., 1992) using Bushnell-Hass mineral broth (Bushnell and Hass, 1941) with *n*-hexadecane as the sole carbon source.

A five-tube MPN technique was used. Three sets of five screw-cap tubes, each containing 5 ml of Bushnell-Hass mineral broth were inoculated with 1, 0.1, and 0.01 ml of a sample dilution.

After inoculation, 50 μL of UV-sterilized *n*-hexadecane (Merck) and 100 μL of autoclaved resazurin solution (Merck, 50 mg/L) were added to each tube.

After incubation at 30°C for 7 days, tubes giving positive reaction were counted. The color of positive tubes changed from blue to pink.

TABLE 5. TAHB count at the beginning of the trial.

Container N°	CFU/g
1	6.5×10^5
2	0
3	1.2×10^6
4	3.2×10^6
5	6.7×10^5
6	4.3×10^6
7	4.8×10^6
8	6.5×10^6
9	3.0×10^6
10	1.3×10^6
11	1.7×10^6

TABLE 6. HDM count at the beginning of the trial.

Container N°	MPN/g
1	2.4×10^8
2	0
3	4.6×10^8
4	6.4×10^7
5	4.6×10^8
6	$> 1.1 \times 10^9$
7	1.1×10^9
8	4.6×10^8
9	$> 1.1 \times 10^9$
10	2.4×10^8
11	$> 1.1 \times 10^9$

RESULTS AND DISCUSSION

In 90 days the moisture varied between 5 and 15%. The pH varied between 4.7 and 5.3.

TABLE 7. General results for each one containers.

Container N°	Strain inoculated	Fertilizer	Total % (w/w) of hydrocarbons reduction	% (w/w) partial of hydrocarbons reduction obtained:		
				30 days	60 days	90 days
1	-	-	12	4	6	2
2	-	-	8	4	2	2
3	W1	-	34	26	-2	12
4	W2	-	43	26	16	9
5	W3	-	22	13	1	10
6	W1	Z1	31	6	23	5
7	W1	Z2	38	29	5	8
8	W2	Z1	26	19	14	-6
9	W2	Z2	45	33	6	13
10	W3	Z1	33	17	5	15
11	W3	Z2	21	10	-2	15

TAHB count (average value for 90 days): 2.6×10^6 CFU/g. HDM count (average value for 90 days): 8.8×10^7 MPN/g. In most containers, the highest percentages of hydrocarbons reduction were observed within the first 30 days.

The biggest biodegradation efficiency was obtained in the containers N°9, 4 and 7. Container N° 9 exhibited the biggest reduction 45%.

CONCLUSIONS

W2 strain is the one which seems to possess the greater capacity of biodegradation.

Z2 fertilizer has been more effective than Z1 fertilizer; except in those containers in which the W3 strain was inoculated.

The biodegradation process for the W2 strain depends of the phosphorus contribution, which would be one limiting nutrients of the process. Fertilization stimulated the biodegradation process. This strain could be used to accelerate the processes of in-situ bioremediation of contaminated soils with hydrocarbons.

ACKNOWLEDGEMENTS

The author thanks to the authorities of the Faculty and Refinery San Lorenzo that believed in the potentials of this project and supported it.

REFERENCES

Atlas, R. M., and R. Bartha. 1992. "Hydrocarbon Biodegradation and oil spill bioremediation". In K. C. Marshall (Ed.), *Advances in Microbial Ecology, 12*: 287-338, Plenum Press, N.Y.

Belloso, C. O., J. Carrario, and D. Viduzzi. 1997. "Hydrocarbons degrading-bacteria isolated of contaminated soils with hydrocarbons" (in Spanish). In *Proceeding III Jornadas Rioplatenses de Microbiología*. Asociación Argentina de Microbiología. pp. 76, E9. Buenos Aires, Argentina.

Bushnell L. D., and H. F. Hass. 1941. "The utilization of certain hydrocarbons by microorganisms". *J. Bacteriol. 41*: 653-673.

Clesceri, L. S., A. E. Greenberg, and R. Trussel Rhodes (Eds.). 1992. "Standards Methods for the Examination of Water and Wastewater". 18th ed., American Public Health Association, N.Y.

Ferrari, M. D., C. Albornoz, and E. Neirotti. 1994. "Biodegradability on soil of residual hydrocarbons from petroleum tank bottom sludges" (in Spanish). *Revista Argentina de Microbiología, 26*: 157-170.

Leahy, J. G., and R. R. Colwell. 1990. "Microbial degradation of hydrocarbons in the environment". *Microbiol. Rev. 54*: 305-315.

OPTIMAL DESIGN AND OPERATION OF LAND-TREATMENT SYSTEMS FOR PETROLEUM HYDROCARBONS

Sedat Kıvanç (Middle East Technical University, Ankara, Turkey)
Kahraman Ünlü (Middle East Technical University, Ankara, Turkey)

ABSTRACT: Land treatment technology has been extensively used for the disposal of petroleum hydrocarbon containing wastes. Effective management of land treatment systems requires optimum·design and operation of the system in order to achieve the fastest and most complete degradation of petroleum hydrocarbons without contamination of the environment. This paper describes a model that can be used for optimal design and operation of land treatment systems for petroleum hydrocarbon containing wastes. The model is composed of system simulator and optimization submodels. Selected model applications are presented to demonstrate the applicability and utility of the model. Such model applications include determination of the optimal design and operation conditions for the land treatment of petroleum hydrocarbon containing wastes under various different site and soil environmental conditions and practical waste disposal scenarios.

INTRODUCTION

If managed properly, land treatment is a viable and cost-effective technology for hazardous waste disposal. This technology has been widely used for the disposal of petroleum hydrocarbon containing wastes due to its remarkable advantages such as less energy requirement, reduced long-term liabilities and low initial and operational costs compared to other alternatives. During the land treatment, waste material is incorporated in biologically the most active upper zone of the surface soils, which serves as a treatment medium for the degradation, transformation or immobilization of the hazardous constituents in the waste.

Optimal design and operation conditions for a land treatment system can be controlled to a large extent by natural biological, chemical and physical soil processes, soil environmental characteristics, site and waste characteristics. Improving the effectiveness of land treatment technology via proper design and optimal operation requires an understanding of these various controlling processes as well as their interactions. Mathematical models can be considered as a powerful tool for such an understanding and application of land treatment systems for petroleum hydrocarbon containing wastes.

Objective. The objective of this study is to present the development and practical applications of a mathematical model that can be used as an effective tool for optimal design and operation of land treatment systems under various different site characteristics and disposal scenarios for hydrocarbon containing wastes.

METHODS

The proposed model is composed of system simulation and optimization submodels. Conceptually, the system simulation submodel is composed of "waste mixing zone" and "lower treatment zone" modules (Short, 1985). Figure 1 shows a conceptualized schematic profile of a land treatment system. The system simulation submodel allows for periodic waste applications and determines the spatial and temporal variation of the state variables such as water content, phase summed (total) and aqueous phase contaminant concentrations in the system. The optimization submodel coupled with the system simulation submodel determines the optimal values of the system control variables, such as waste loading rate, infiltration rate, water content, frequency of waste application and the dimensions of the land treatment system under the constraints of satisfying a prescribed water quality criteria in the aquifer.

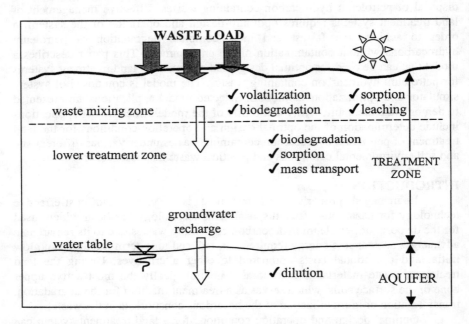

FIGURE 1. Conceptual representation of a land-treatment system

The waste mixing zone was treated as a "complete-mix reactor" using a mass balance expression for the phase summed total mass of hydrocarbon components present in the soil-waste mixture. The phase-summed concentration is expressed as:

$$C_T = C_{ks} + \theta_w C_{kw} + \theta_a C_{ka} + \theta_o C_{ko} \tag{1}$$

where C_T = total concentration of contaminant in the waste zone (g/cm^3)
θ_w = volumetric water content (cm^3/cm^3)
θ_a = volumetric air content (cm^3/cm^3)
θ_o = volumetric hydrocarbon content (cm^3/cm^3)
C_{ks} = solid phase concentration of contaminant species k (g/cm^3)
C_{kw} = aqueous phase concentration of contaminant species k (g/cm^3)
C_{ka}, = air phase concentration of contaminant species k (g/cm^3)
C_{ko} = hydrocarbon phase concentration of contaminant species k (g/cm^3)

Linear relationships were assumed to describe the partitioning between the phases.

Depletion of the hydrocarbon source over time was assumed to occur due to combined effects of leaching by water percolating through the waste zone; diffusive losses from the soil surface for volatile components; and biodegradation losses due to microbial activities. The aqueous loss is purely advective and mass flux density in the water phase is directly related to water flux (Darcy's velocity) and the mass of contaminants per volume of aqueous phase. Volatilization losses were modeled using the approach by Thibodeaux (1981). Biodegradation in the soil matrix was assumed to follow first order kinetics through the treatment zone. It was assumed that biological activity is the highest within the waste mixing zone, resulting in a maximum decay rate constant.

A mass balance equation yields total concentration of contaminant in the waste mixing zone as a function of time as follows:

$$C_T = C_T^o \cdot \exp(-\beta_{kT} \cdot t) \tag{2}$$

where C_T^o = initial total contaminant concentration (g/cm^3)

β_{kT} = overall mass depletion coefficient for the contaminant k (1/day). The initial contaminant concentration is described by

$$C_T^o = \frac{F_H \cdot f_k^o \cdot \rho_k \cdot \rho_b}{\rho_o} \tag{3}$$

where F_H = mass fraction of hydrocarbon in the waste mixing zone (g/g)

f_k^o = volume fraction of contaminant k in the hydrocarbon phase (cm^3/cm^3)

ρ_k = specific density of the contaminant k (g/cm^3)

ρ_b = waste zone bulk density (g/cm^3)

ρ_o = average density of hydrocarbon.

In *the lower treatment zone*, the mass balance was written in terms of the aqueous phase concentration. Contaminant transport is modeled from bottom of the waste mixing zone to the aquifer assuming negligible horizontal spreading. Vertical one-dimensional, unit gradient unsaturated leachate flow was assumed at a moisture content controlled by the net water flux and the unsaturated hydraulic conductivity function. Contaminant transport through the lower treatment zone into the water table occurs by advection and dispersion with linear adsorption and first-order decay. The biodecay rate constant exponentially decreases with depth away from the waste mixing zone.

For the case of single waste load applications an analytical solution to the advective-dispersive equation was used. For the case of multiple periodic waste load application, the surface boundary condition associated with the waste mixing zone can be given as

$$C_{kw}(0,t) = \sum_{i=1}^{N} C_{kw}^o \exp(-\beta_{kT}t) \quad 0 \le t \le T_i \tag{4-a}$$

$$C_{kw}(0,t) = 0 \quad t \ge T_i \tag{4-b}$$

where C_{kw}^{o} = initial aqueous concentration of hydrocarbon species k (g/cm^3)

N = the number of periodic waste applications

T_i = the time between the two consecutive waste load applications

i = index for individual waste application, i.e., $i = 1,2,\ldots,N$

With this boundary condition, no longer an analytical solution for the transport equation is available. Thus, a finite difference numerical solution of the 1-D transport equation is obtained to calculate the contaminant concentration within the lower treatment zone including the water table in the aquifer.

For optimization, the objective function is formulated as maximizing the mass losses or equally minimizing the mass accumulation in the waste mixing zone. Explicitly the objective function, with the terms occurring at the ith waste load application, can be written as

$$\text{Min} \sum_{i=1}^{N} MT_i - \left[ML_i + MV_i + MB_i \right] \tag{5}$$

where MT_i = total mass added to the land treatment system (g)

ML_i = total mass leached from the system (g)

MV_i = total mass volatilized from the system (g)

MB_i = total mass biodegraded from the system (g)

Two constraints are stated within the optimization submodel:

$$f\left(C_{aq}, F_H, q_w, L_w, W, T \right) = C_{MCL} \tag{6-a}$$

$$\theta_w + \theta_o + \theta_a = \phi \tag{6-b}$$

Where F_H = waste load application rate (g/g)

q_w = net water flux through the treatment zone (cm/day)

L_w = thickness of the waste mixing zone (cm)

W = width of waste mixing area (cm)

T = frequency of waste load applications

Equation 6-a describes the aqueous phase contaminant concentration in the aquifer, C_{aq}, as a function of the system control variables F_H, q_w, L_w, W and T. C_{aq} should not exceed the allowed maximum concentration level (C_{MCL}).

Equation 6-b dictates that the sum of the volumetric contents of water, air and hydrocarbon phases is equal to the total porosity, ϕ, of the soil system. The volumetric water content is implicitly a function of water flux and hydraulic conductivity, also hydrocarbon content is a function of application rate of waste.

The above optimization problem has mathematically nonlinear objective function and nonlinear constraint functions. The model minimizes the objective function while satisfying the constraints by computing the system state variables; and, determines the optimal values of system control variables for the optimum management of land treatment systems.

Two hypothetical but realistic operation scenarios were proposed in order to illustrate the optimal design and operation of land treatment systems. First one is over a fixed period of operation time, e.g. 5 years, determining the optimal amount of waste that can be disposed of; and second one is, for disposing of a fixed total amount of oily waste, e.g.,100 kg/m^2, determining the optimal duration of operating the land treatment system.

RESULTS AND DISCUSSION

To demonstrate the developed model, various simulation runs are performed using single and multiple waste load applications. Simulation runs involve demonstrating the performance of a land treatment system for oily wastes at sites with loam, clay and sand soil, representing a wide spectrum of hydrologic regimes. An aromatic compound, benzene, is selected as the index compound based on the fact that benzene is a high priority pollutant for most water quality concern. Base-case parameters of soil and site characteristics were obtained from Ünlü et al., 1992; Newell et al., 1990; Carsel and Parrish, 1988.

Normalized sensitivity analyses results showed that some parameters are highly sensitive to the model while some are less sensitive. Among more than twenty parameters, organic carbon mass fraction, porosity and vanGenuchten moisture retention parameter seem to be most sensitive parameters in the waste mixing zone. On the other hand, water flux density shows relatively high and asymmetric sensitivity distribution for the lower treatment zone.

Single and multiple waste load simulations show that mass losses due to biodegradation are dominant for each of the three types of soil; loam, clay and sand soils. Biodegradation is responsible for about 90% of total mass fluxes in loam and clay soils and 60% for sand soil. Volatilization losses has little effect in clay soil and has nearly 30% contribution for sand soil. In the case of multiple waste loadings, attenuation characteristic of the soil media is clearly observed especially at shallow depths.

Optimized system control variables for scenario 1 and 2 are given in Table 1. N, the total number of waste loadings, and θ_w, volumetric water content are the parameters which were implicitly optimized.

TABLE 1. Optimal values of system control variables for scenarios 1 and 2.

	Scenario 1			Scenario 2		
	Loam	Sand	Clay	Loam	Sand	Clay
F_H, (g/g)	0.0300	0.0487	0.0200	0.0703	0.0300	0.0335
q_w, (cm/d)	0.287	0.25	0.153	0.309	0.258	0.179
L_w, (cm)	15	15	15	15	32	35
W, (cm)	1864	3000	1375	1000	3000	3000
T, (d)	300	46	30	64	75	80
N	6	40	61	4	9	8
θ_w, (cm^3/cm^3)	0.2669	0.1178	0.3008	0.2689	0.1184	0.3041
Total amount of waste disposed, (kg/m^2)	40.8	441.2	286.0	-	-	-
Total operation time (d)	-	-	-	256	675	640
Value of objective function (mg/m^2)	6.688x 10^{-4}	8.304x 10^{-14}	533.6	639.0	3.347x 10^{-6}	1104.0

The minimized values of objective function given in Table 1 indicates that mass accumulation in sand soil much less than in loam and clay soils. So, sites with sand soils exhibit preferable performance for waste treatment under the considered hydrogeologic settings. Examining the results of the second scenario, loam soil appears to be the most efficient among three soil types. From the operational stand point, scenario 1 implies the situations where the time is limiting factor on waste disposal activities; however, scenario 2 implies situations where the total amount of waste to be disposed of is a limiting factor rather than operation time. Model results related to be the numerical values of objective function also concur with the well known fact that clay soils are well suited media for waste containment rather than waste treatment.

CONCLUSIONS

This study proposes a modeling approach that combines the system simulation and optimal system control concepts, in order to address the problem of optimal management of land treatment systems which lately has been gaining a growing interest from the industry. The developed model has a sound theoretical background since it is developed based on analytical tools that are widely accepted by the literature. The model has the capabilities of predicting fundamental state and control variables of the land treatment system. Thus, model results can be used for the optimal design and operation of land treatment system under a wide range of site characteristics and waste disposal scenarios. The model also has a flexible structure allowing for the user to modify the model to account for different environmental processes, contaminants and management scenarios.

REFRENCES

Carsel, R. F. and R. S. Parrish. 1988. "Developing joint probability distributions of soil water retention characteristics." *Water Resour. Res.* 24(5): 755-769.

Newell, C. J., L.P. Hopkins, and P.B. Bedient. 1990. "A hydrogeologic database for ground-water modeling." *Ground Water* 28(5): 703-714.

Short, T. E. 1985. "Movement of contaminants from oily wastes during land treatment." in (Eds) E. J. Calabrese and P. T. Kostecki, *Soils Contaminated by Petroleum*.

Thibodeaux, L.J. 1981. "Estimating the air emissions of chemicals from hazardous waste landfills." *Journal of Hazardous Materials* 4: 235-244.

Ünlü, K., M. W. Kemblowski, J. C. Parker, D. Stevens, P. K. Chong, I. Kamil. 1992. "A screening model for effects of land-disposed wastes on groundwater quality." *Journal of Contaminant Hydrology* 11: 27-49.

BIOREMEDIATING PAHs AND TPH AT THE WATERVLIET ARSENAL

J. W. Talley, USAE Waterways Experiment Station, Vicksburg, MS, USA
Kenneth J. Goldstein, Ralph G. Schaar, Malcolm Pirnie Inc., Mawah, NJ, USA
Paul B. Hatzinger, Envirogen Inc., Lawrenceville, NJ, USA
Sanjib Chaki, USACE-Baltimore District, Baltimore, MD, USA
(currently with USEPA, Washington DC)
Maira Senick, Watervliet Arsenal, Watervliet, NY, USA

ABSTRACT: Bench-scale laboratory treatability studies on contaminated soils were initiated to (1) provide evidence of biodegradation of total petroleum hydrocarbons (TPH) and polycyclic aromatic hydrocarbons (PAHs), and (2) assess the potential for PAH degradation under either aerobic and/or anaerobic conditions with and without the addition of nitrate as an electron acceptor. The studies consisted of a 9-week microcosm experiment using ^{14}C-phenanthrene, a 9-month aerobic soil column study, and a series of toxicity bioassay analyses. Bacterial isolation, bioslurry, and pan studies are currently underway. PAH-mineralizing bacteria were found to be present and active, resulting in a PAH reduction of 40-58% over the 9-month study, depending on the specific treatment. TPH biodegradation exceeded 73% in every treatment. Controls suggested that mass transfer limitations (oxygen and nutrients) could be controlling both TPH and PAH degradation. Toxicity was reduced after biotreatment in every case. Overall, the data from this work suggest that landfarming (to increase oxygen transfer) with addition of inorganic nutrients may be an appropriate treatment option at this site.

INTRODUCTION

Watervliet Arsenal (WVA), located in Watervliet, New York, is the United States' oldest, continuously operating, cannon manufacturing facility. The City of Watervliet is situated near the eastern boundary of Albany County, New York, on the West Bank of the Hudson River, approximately four miles south-southwest of the confluence of the Mohawk and Hudson Rivers. The arsenal consists of two contiguous areas that comprise a total land area of 140 acres. The main process area of the arsenal (Main Manufacturing Area) is a 125-acre tract on which manufacturing and administrative operations occur.

The Siberia Area, a 15-acre tract of land, is used as a shipping yard and for the interim storage of raw and hazardous materials, finished goods and supplies for the Arsenal. A Resource Conservation and Recovery Act (RCRA) Facility Investigation (RFI) of the WVA Main Manufacturing Area and Siberia Area has been performed. The results of the RFI for the Siberia Area identified polycyclic aromatic hydrocarbons (PAHs), petroleum hydrocarbons (i.e., diesel fuel, kerosene and lubricants) and heavy metals (specifically chromium, lead, and arsenic) as the primary contaminants in the soil.

The major sources of the PAH and petroleum hydrocarbon contamination in the Siberia Area are waste oil, which was applied to the ground to control dust, and oil-saturated metal chips resulting from manufacturing processes at the site. PAHs ranged from very low levels (1-30 mg/kg) to hot spot values as high as 325,000 mg/kg. TPH concentrations in soils averaged approximately 2,000 mg/kg, but exceeded 45,000 mg/kg in some locations.

Objective. The objective of this study is to determine the feasibility of bioremediation for the treatment of PAH and TPH impacted soils at the Siberia Area, Watervliet Arsenal. Specific goals are to (1) determine the capability of indigenous organisms to aerobically and anaerobically biodegrade PAHs; (2) isolate PAH-degrading bacteria; (3) demonstrate TPH and PAH biodegradation in soil slurries; and (4) assess soil toxicity before and after bioremediation using a Microtox bioassay. To achieve this objective, a series of laboratory experiments consisting of a 9-week microcosm experiment with ^{14}C-phenanthrene, a 9-month aerobic soil column study, and a series of toxicity bioassay analyses were performed. Bacterial isolation, bioslurry and pan studies are currently underway.

MATERIALS AND METHODS

Sample Collection. Soil samples were collected from the Siberia Area at locations determined to be "representative" of the PAH and TPH contamination at the site. A clean backhoe was used to break the surface and expose contaminated

Watervliet Arsenal

FIGURE 1. Soil Samples from Siberia Area, Watervliet Arsenal

soil down to approximately 3 feet. Soil samples were collected with stainless steel trowels and placed in non-reactive plastic containers, which were completely filled to minimize headspace. Samples were maintained at 4°C and shipped within 24 hours to Malcolm Pirnie's Environmental Technology Laboratory in Monsey, New York and Envirogen's Treatability Laboratory in Lawrenceville,

New Jersey. Samples were initially passed through a wire sieve (6.7 mm mesh size) to remove rocks and debris, then they were stored in tightly sealed inert containers at 4°C until use.

[14]C-Phenanthrene Study. Twenty-five grams of soil (dry weight) was added to 15 autoclaved serum bottles (160 ml, Fisher Scientific). Twelve of the sample bottles were prepared in an aerobic chamber (Coy Laboratory Products, Inc) with a headspace of approximately 95 % N_2 and 5 % H_2 gas. The other 3 bottles were prepared under aerobic conditions.

FIGURE 2. Microcosms used in the ^{14}C-Phenanthrene Study

Each soil sample was amended with 10 µl of a solution containing 12.5 µg of unlabeled phenanthrene (98 % purity, Aldrich Chemical Co.) and 300,000 dpm of [9-^{14}C] phenanthrene (46.9 mCi/mmol, Sigma Chemical Co.) prepared in dichloromethane. The phenanthrene stock was added dropwise at several locations in the soil sample, then the soil was vigorously shaken several times over a 30 minute period to distribute the phenanthrene and to allow the dichloromethane to volatilize from the soil. Triplicate bottles prepared under anaerobic conditions were amended with 1.0 ml of deoxygenated water containing the following: (1) nitrate (2.25 mg KNO_3); (2) nitrate (2.25 mg KNO_3) and inorganic nutrients [19.2 mg $(NH_4)_2HPO_4$ and 9.6 mg $(NH_4)H_2PO_4$]; (3) no addition (distilled water only); or (4) $HgCl_2$ and NaN_3 (12.5 mg each) to inhibit biological activity. The three samples prepared aerobically were amended with 1.0 ml of water with inorganic nutrients [19.2 mg $(NH_4)_2HPO_4$ and 9.6 mg $(NH_4)H_2PO_4$]. The addition of 1.0 ml of water to each soil sample brought the moisture level to approximately 80 % of the measured water-holding capacity for the soil.

All serum bottles were crimp-sealed with Teflon-lined serum stoppers. The samples were incubated in the dark at room temperature (23 $\pm$ 3°C).

Periodically, the headspace in the anaerobic samples was flushed for 12 minutes with N_2 gas, which was initially passed through a heated column of copper filings to remove trace levels of oxygen. The aerobic samples were flushed with air. The headspace gases exiting each bottle were passed through a sparging train consisting of two scintillation vials containing Liquiscint scintillation cocktail (National Diagnostics) followed by two vials containing Oxysol ^{14}C-scintillation cocktail (National Diagnostics). The first two vials trap ^{14}C-volatile organics from the headspace and the latter two vials trap [14]CO_2. For a detailed description of this flushing technique and system, see Marinucci and Bartha (1979).

The quantities of ^{14}C-volatiles and [14]CO_2 in the headspace of each bottle were quantified by enumerating the ^{14}C in each trapping vial. At each flushing time, the quantity of ^{14}C in the first two vials was summed (^{14}C-volatiles in headspace) and that in the last two vials was summed ([14]CO_2 in headspace). The extent of mineralization in samples was determined from the cumulative amount of [14]CO_2 collected from each sample.

Soil Column Study. Soil samples were combined then vigorously mixed for 10 minutes using solvent and acid washer implements. After homogenization, a baseline concentration for TPH, PAHs, total organic carbon (TOC), ammonia, nitrate, and orthophosphate was established using appropriate analytical methods. Initial values for TPH and total PAHs were 1,930 mg/kg and 8 mg/kg, respectively. These samples had lower PAH concentrations than expected; nonetheless, a comparison between these values and New York State Department of Environmental Conservation Technical and Administration Guidance Memorandum Target Cleanup (TAGM) Levels revealed that some PAHs (namely benzo(a)anthracene, chrysene, benzo(a)pyrene, and dibenz(a,h)anthracene) were still approximately an order of magnitude above appropriate levels.

The homogenized soil was placed into four, 4-inch diameter, 30-inch tall plexiglas columns with sampling ports placed at heights of 6, 12, and 18 inches. The columns were maintained in a laboratory hood at approximately 25°C. The columns consisted of two controls (Controls A and B) and two treatment columns (Treatments A and B). The soil in Control A and Treatments A and B was mixed three times a week in an inert cooler then repacked into columns. Control B was not mixed. The water was adjusted in all columns to maintain soil moisture levels of 15% to 20%. Carbon substrate ($C_6H_{12}O_6$), nitrogen (NH_4NO_3), phosphorus (KH_2PO_4) were added to Treatments A and B at a ratio of 100:10:1 (mol/mol/mol). Nutrient measurements were taken every ten days using a Hach NPK-1 Soil Analysis Kit. Nutrient amendments were performed when nitrogen and phosphorus levels fell below 25% of the added amount. O_2 and CO_2 levels in soil columns were measured once a week using Draeger tubes.

Each of the four columns underwent six sampling events for laboratory chemical analysis over the 9-month study. The first three samples were taken monthly during the initial three months of incubation. The final three samples were taken every two months during the final six months of the study. PAHs were analyzed in each sample. TPH was analyzed only during the last three sampling events.

FIGURE 3. Soil Column Study Treatments

Toxicological Bioassays. Toxicological bioassays were conducted on background soil, contaminated soil, and biotreated soil. The background soil was collected from a golf course located within Watervliet Arsenal. The contaminated soil was collected from a heavily contaminated area within Siberia (substation area). The biotreated samples were collected from Treatments A and B at the conclusion of the 9-month study. All soil samples were placed in 150-mL sample jars, cooled, and submitted to an outside laboratory for Microtox bioassays. During Microtox analysis, samples are exposed to an aqueous suspension of luminescent bacteria (Vibrio fisher). Toxic materials present in the test sample interfere with the luminescence of the bacteria, resulting in a reduction of light output. The light output is recorded at the level in which they correspond to 50% toxicity thresholds for the test bacteria. These measurements are referred to as EC_{50} values. Lower EC_{50} values, compared to background levels, indicate increased toxicity.

RESULTS

^{14}C-phenanthrene Study Results. The mineralization of ^{14}C-phenanthrene to $^{14}CO_2$ for the aerobic and anaerobic treatments is plotted as a function of incubation time in Figure 4. The error bars represent the standard deviations from triplicate samples. Rapid and extensive mineralization of ^{14}C-phenanthrene was observed in the aerobic samples amended with nutrients. In these soil samples, more than 30 % of the added ^{14}C-phenanthrene was mineralized to $^{14}CO_2$ after 5 days, and greater than 60 % was collected as $^{14}CO_2$ after 64 days of incubation. Appreciable mineralization of phenanthrene was also observed in all anaerobic samples except for those treated with the biological inhibitors $HgCl_2$ and NaN_3 (killed controls). More than 13 % of the added phenanthrene was detected as $^{14}CO_2$ within 5 days in all anaerobic treatments. After 64 days, 34%, 37%, and 44 % of added phenanthrene was collected as $^{14}CO_2$ in anaerobic samples amended with water, nitrate, and nitrate plus nutrients, respectively. The quantity of ^{14}C-volatiles during headspace flushing was trivial in all treatments (< 0.5 % of the added ^{14}C-phenanthrene).

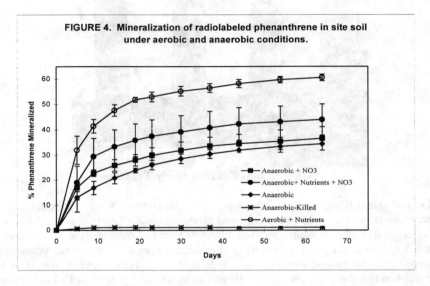

FIGURE 4. Mineralization of radiolabeled phenanthrene in site soil under aerobic and anaerobic conditions.

Soil Column Study Results. Total PAH reductions were noticeable in those columns that maintained mixing and nutrients. After two months of mixing, total PAH concentrations increased significantly. This phenomenon has been observed in several studies involving hydrophobic organic compounds (Hatzinger, P.B. 1996; Alcoa, 1994). These "high" values were compared to concentrations present after 9 months of treatment. These values were used to calculate the percent rate of reduction. Both Treatments A and B showed reduction in total PAHs. Treatment A had a 40.77% reduction based on a high concentration value of 12.47 mg/kg. Treatment B had a 58.16% reduction based on a high concentration value of 26.94 mg/kg. Control A had very little biodegradation (8.03%) of total PAHs. Control B had no biodegradation (0.0%) of total PAHs. Using the initial TPH values, a 73.58% reduction in Treatment A and a 76.48% reduction in Treatment B were observed. Control A shows only a 30.05% reduction in TPH. Control B shows a TPH reduction of 69.79%.

Toxicity Bioassay Results. EC_{50} values for background samples ranged from 3.34% to 9.76% with an average value (95% Confidence Interval) of 5.80%. Initial EC_{50} values for the contaminated soil samples ranged from 1.61% to 2.37% with an average value of 1.95%. EC_{50} values for the biotreated samples

(Treatments A and B) were only slightly higher than pretreatment values. EC_{50} values ranged from 1.85% to 2.30% (average of 2.06%) for Treatment A and 2.45% to 3.87% (average of 3.08%) for Treatment B. EC_{50} values for Control A ranged from 1.52% to 2.16% (average of 1.81%). EC_{50} values for Control B ranged from 1.31% to 1.88% (average of 1.57%). An additional contaminated soil sample was collected from a known hot spot for future integration into the study. It exhibited EC_{50} values ranging from 0.0148% to 0.0204% with an average value of 0.0174%. This sample represents a 100-fold increase in toxicity.

DISCUSSION

Hydrophobic (water-hating) organic compounds (HOCs) such as PAHs have a high affinity for sorption onto soil particles and into soil aggregates. This characteristic limits the availability of the contaminant to the microbial community. Aggressive mixing can help to reduce this mass transfer limitation by breaking down the soil aggregates and increasing the physical contact between water present in the soil and the PAHs. As more PAHs are solubilized into the aqueous phase, bioavailability is generally increased. Mixing also serves to enhance the transfer of oxygen (electron acceptors) within the soil while homogenizing the soil with respect to other contaminants (TPH) and amendments. Although TPHs are not highly hydrophobic like PAHs, they too can show an initial increase in availability. If an initial increase in TPH occurs, it is normally observed within the first 30-60 days of mixing. The overall effect of mixing generally results in increased bioavailability and enhanced biodegradation. Observing biodegradation trends associated with the "controls" and "treatments" can provide critical information about assessing the potential success for engineered biotreatment.

The ^{14}C-phenanthrene study indicated that aerobic and anaerobic PAH degraders are present at the site. Rapid initial biodegradation rates occurred in both aerobic and anaerobic samples suggesting that both processes may be useful for bioremediation. However, greater extent of mineralization in aerobic compared to anaerobic samples suggests that aerobic processes are more efficient. Aerobic biodegradation of PAHs with less than five rings has been widely reported, however, PAHs have traditionally been considered resistant to biological attack under anaerobic conditions. A few studies have revealed the potential for PAH biodegradation under denitrifying (Mihelcic and Luthy, 1988) and sulfate-reducing (Zhang and Young, 1997) conditions.

In the soil columns, it is important to understand that biological systems can respond differently from one analysis to another. This coupled with the specific heterogeneity of HOCs in the Siberia Area can result in significant variances in chemical data. One method of minimizing these effects is to perform triplicate analysis of samples and then average the results. This provides data that is more statistical representational for interpretation. Due to budgetary limitations this work utilized only single analysis. Therefore, the focus was on observing general trends associated with the data. In other words, do concentrations over time decrease or stay the same.

Treatments A and B represent the level of engineered biotreatment associated with in-situ landfarming. Since Treatment B had over twice the amount of PAHs present, it is possible that PAHs were less limiting (more available) than in Treatment A. This could be one possible explanation for the increased rate of biodegradation. This seems to suggest the importance of having nutrients added, especially when mixed wastes (TPH and PAHs) are present. Little, if any, biodegradation occurred in Controls A and B indicating the significance of having aggressive mixing to overcome mass transfer limitations. In essence, Control B also shows that natural attenuation or non-engineered biotreatment scenarios cannot be effective in remediating PAHs.

Both Treatments A and B were effective in reducing the concentration of TPH (greater than 75%). In contrast, Control A showed only a 30.05% reduction in TPH, indicating that nutrients may have become limiting as aerobic bioactivity increased with aggressive mixing. Control B represents the effects of natural attenuation or intrinsic bioremediation. Since natural attenuation of TPH is well documented, some biodegradation in Control B should be expected, albeit at much slower rates. Control B shows a TPH reduction of 69.79%. This suggests that anaerobic or anoxic natural attenuation of TPH is occurring, albeit at a slightly lower rate than that of the aerobic biodegradation of Treatment A and B. It also supports the premise that mass transfer limitations do not generally control the availability of TPHs.

Toxicity bioassays using the Microtox technique were also performed on the soil before and after biotreatment. Comparisons of EC_{50} values of the biotreated soils with background (clean) soil were made in an attempt to relate bioavailability and toxicity for purposes of supporting alternative environmental endpoints. However, the results from this comparison were inconclusive. While Microtox analyses may be a good screening tool to differentiate between high concentrations (i.e. hot spots) and low concentrations, this approach appears to fall short of assessing gradations of contaminant levels. In addition, Microtox analyses may not be sensitive enough to detect the effects of hydrophobic compounds such as PAHs. Therefore, the toxicity assessment effort of this study was refocused to include earthworm bioassays. This work is currently underway using the additional soil samples, which exhibited high toxicity (low EC_{50} values).

Overall, the results of the 9-month soil column study in conjunction with the rapid and extensive mineralization of ^{14}C-phenanthrene observed under aerobic conditions, without an apparent lag phase, suggests that landfarming (to increase oxygen transfer) with addition of inorganic nutrients may be an appropriate treatment option at this site. A better understanding of PAH degradation under anaerobic conditions could lead to new treatment technologies for accelerated PAH bioremediation utilizing alternative electron acceptors, such as nitrate or sulfate. Completion of the current bioslurry and pan studies utilizing the new collected samples (from hot spots) should address these remaining issues; as well as, provide scale-up parameters for field pilot demonstration. It is expected that this work will improve the decision-making process about the fate of contaminants in soils, reduce treatment costs associated with clean up, and increase the potential for the reuse of treated contaminated soils.

REFERENCES

ALCOA, 1994. *Final Report, Evaluation of In Situ Bioremediation Treatment of 60-Acre Lagoon Sludge Material for PCB Reduction/Immobilization at Alcoa's Massena, New York Facility.* Aluminum Company of America, Massena, NY.

Hatzinger, P.B. 1996. "Effects of Aging on the Biodegradation of Chemicals." Ph.D. Dissertation, Cornell University, NY

Marinucci, A. C. and R. Bartha. 1979. "Apparatus for monitoring the mineralization of volatile ^{14}C-labeled compounds." *Appl. Environ. Microbiol.* 38:1020-1022.

Mihelcic, J. R., and R. G. Luthy. 1988. "Degradation of polycyclic aromatic hydrocarbon compounds under various redox conditions in a soil-water system." *Appl. Environ. Microbiol.* 54:1182-1187.

US Army, 1997. *Draft Corrective Measures Study (CMS) Field Data Report, Siberia Area.* Watervliet Arsenal, Watervliet, NY.

US Army, 1997. *Final RCRA Facility Investigation Report, Siberia Area.* Watervliet Arsenal, Watervliet NY.

Zhang, X., and L. Y. Young. 1997. "Carboxylation as an initial reaction in the anaerobic metabolism of naphthalene and and phenanthrene by sulfidogenic consortia." *Appl. Environ. Microbiol.* 63:4759-4764.

REGULATORY AND MANAGEMENT ISSUES IN PREPARED BED LAND TREATMENT: LIBBY GROUNDWATER SITE

J. Karl C. Nieman (Utah State University, Logan, Utah)
Ronald C. Sims (Utah State University, Logan , Utah)
David M. Cosgriff (Champion International, Libby, Montana)

ABSTRACT: Prepared bed land treatment has been used to treat creosote and pentachlorophenol contaminated soil at the Libby Groundwater Site since 1989. Since the initiation of treatment, several changes in the remedial standards applied to the site have been approved. Changes in management of the treatment process have also been implemented or investigated. Effects of regulatory and management changes are discussed with respect to their impact or potential impact on treatment time and site closure. Results are applicable to the continued management of the Libby site and to other sites using prepared bed land treatment to treat wood-preservative contaminated soils.

INTRODUCTION

Prepared bed land treatment has been a major bioremedial treatment technology used in the treatment of wood-preservative contaminated soils over the past decade. The Libby Groundwater Site in Libby, Montana was one of the first listed Superfund sites in 1983 and has employed this technology in the treatment of wood-preservative contaminated soil since 1989.

A bioremediation field performance evaluation conducted during 1991 and 1992 concluded that the biological treatment process at the site resulted in reduction of soil contaminant levels to specified standards and detoxification of contaminated soil (U.S. EPA, 1996). Although the treatment process was found to be effective, the rate of biological treatment has been a primary concern with respect to the estimated date of site closure. The site initially contained approximately 45,000 bank yds^3 (~34,405 m^3) of soil (particle size < 1 inch [2.54 cm]) contaminated primarily with creosote derived polycyclic aromatic hydrocarbons (PAH) and pentachlorophenol (PCP) due to wood-treating operations between 1946 and 1969. To date approximately 14,000 loose yds^3 (~10,700 m^3) of contaminated soil have been successfully treated at the site. Soil from the site has been the subject of several field and laboratory studies. Based on the results of these studies, some changes have been made in both regulation and management of the remediation process. This paper provides an update of the remedial progress of the Libby site and discusses the effects of changes in both regulatory remediation standards and proposed and implemented management practices on the treatment of the contaminated soil.

REGULATORY EFFECTS ON SOIL TREATMENT TIME

From 1989 to 1998, treatment of contaminated soil has been implemented in two, 1-acre, on-site land treatment units (LTU). Contaminated soil is applied to the LTUs in layers known as "lifts". Each lift is approximately 8 to 10 inches in depth. The lifts are tilled weekly and irrigated periodically to stimulate aerobic biodegradation of the contaminants. A U.S. EPA approved monitoring plan specifies periodic monitoring of PCP (EPA Method 8040) and PAH (EPA method 8100) concentrations within each LTU lift being treated. Monitoring of dioxin and furan levels were also specified, but are not addresses in this report. Each LTU is divided into 4 zones for sampling. A new lift is applied to the LTU when contaminant concentrations in composite samples from each zone have reached preset remedial goals. Remedial goals for PCP and several PAH compounds including pyrene were established in the 1988 Record of Decision (ROD) and were modified in 1993 and 1997 (see Table 1).

TABLE 1. Remedial standards for the land treatment units at the Libby Groundwater Site. Absence of a value indicates that the contaminant was not regulated under the corresponding standard.

Regulated Compound	1988 standard (mg/kg)	1993 standard (mg/kg)	1997 standard (mg/kg)
Pentachlorophenol	37.0	37.0	37.0
Naphthalene	8.0	—	—
Phenanthrene	8.0	—	—
Pyrene	7.3	—	—
Total Carcinogenic PAH	88.0	88.0	—
Benzo(a)Pyrene equivalency	—	—	59.0

Lift treatment time prior to the 1993 regulatory change was primarily controlled by pyrene which was the land limiting constituent (LLC) (the compound upon which remedial decisions are based) for 9 of the 11 treated lifts shown in Table 2. Periods of active lift treatment time (excluding winter months) would have been reduced by 33 and 43%, respectively, had the 1993 or 1997 standards been in place during this time.

Based on the treatment history through 1993, it was expected that the new standard set in 1993, which eliminated pyrene as a regulated contaminant, would result in a significant reduction in treatment time and a resultant cost savings. Table 3 compares lift treatment times under the various regulatory standards for lifts treated from 1993 to 1998 and indicates that because of the recalcitrance of PCP, the reduction in treatment time may not have been as great as was anticipated.

Because lifts listed in Table 3 were treated under the 1993 standard, the actual treatment time based on the 1988 standard is not known for lifts seven and eight of LTU 2 since pyrene concentrations still exceeded the 1988 standard at the end of the treatment period. However, it is reasonable to assume that additional treatment periods for these lifts under the older standard would have been minimal since both lifts had pyrene concentrations of approximately 8.0 mg/kg (1988 regulatory limit=7.3 mg/kg) in the months July and August when treatment was discontinued.

TABLE 2. Comparison of lift treatment times required under the different standards applied to the Libby Site for lifts treated prior to the application of the 1993 standard. Time periods do not include winter months when active treatment was suspended.

| Cell | Lift | date applied | Days of treatment required under corresponding standard | | | LLC[a] |
			1988 std	1993 std	1997 std	
1	1	6-29-89	34	34	32	pyrene
1	2	8-8-89	116	41	0	pyrene
1	3	7-11-90	70	21	21	pyrene
1	4	5-9-91	65	22	22	pyrene
1	5	7-23-91	117	75	75	pyrene
1	6	5-5-92	94	82	33	pyrene
1	7	9-4-92	54	31	31	pyrene
1	8	4-22-93	124	119	119	pyrene
2	1	7-25-91	180	103	103	pyrene
2	2	6-24-92	33	33	33	PCP
2	3	4-20-93	91	91	91	PCP
		Total	**978**	**652**	**560**	
	%	**reduction**[b]	—	**33%**	**43%**	

(a) LLC=land limiting constituent.
(b) "% reduction" indicates the potential reduction in treatment time achieved by application of the new standard when compared to the 1988 standard.

TABLE 3. Comparison of lift treatment times required under the different standards applied to the Libby Site for lifts treated after the application of the 1993 standard. Times do not include winter months when active treatment was suspended.

| Cell | Lift | date applied | Days of treatment required under corresponding standard | | | LLC[a] |
			1988 std	1993 std	1997 std	
1	9	10-5-93	57	57	57	PCP
1	10	6-21-94	342	342	342	PCP
1	11	8-21-96	>301	>301	>301	still being treated
2	4	10-4-93	58	44	44	pyrene
2	5	6-27-94	88	88	88	PCP
2	6	5-10-95	56	56	56	PCP
2	7	7-20-95	>123	123	123	pyrene
2	8	7-16-96	>185	185	185	pyrene
2	9	8-26-97	154	154	154	PCP
		Total	**>1364**	**>1350**	**>1350**	
	%	**reduction**[b]	—	**>1%**	**>1%**	

(a) LLC=land limiting constituent.
(b) "% reduction" indicates the potential reduction in treatment time achieved by application of the new standard when compared to the 1988 standard.

EFFECT OF MANAGMENT PRACTICES ON TREATMENT TIME

Proper management of the bioremediation process is essential for the timely treatment of contaminated soils. As soil treatment has proceeded over the last ten years, several management options have been explored in the field or in the laboratory. Some of these have lead to field implementation at the Libby site. Categories of investigated management practices include soil inoculation or additions of soil amendments, volume reduction by soil screening, treatment under low oxygen conditions, soil mixing, management of environmental factors, and physical changes in the soil treatment facility.

Soil Inoculation and Amendments. Several inoculation or amendment approaches have been tested at the site with the goal of accelerating soil treatment. A 1994 study involved the addition of a "bio-broth" solution that was generated by mixing effluent from on-site bioreactors that had adapted microbial communities capable of degrading PAH and PCP found in the groundwater at the site. Other components of the solution included molasses, ammonium chloride, potassium tripolyphosphate, and fresh dilution water that were mixed and aerated for 18 hours prior to field application. The duration of the study was 2 months during which the bio-broth solution was added to the test cell weekly, while a control cell received unamended irrigation water. Soil testing during the two month period indicated that the addition of the bio-broth solution did not have a significant effect on PCP degradation rates.

Several other soil amendment tests were conducted by groups of students from Libby High School in 1994. The laboratory microcosm studies tested the effectiveness of a strain of White Rot fungus (*Phanerochaete spp.*), a commercial enzyme solution, and a commercial surfactant solution intended to increase contaminant bioavailability. All three soil amendments indicated some potential for increased degradation of PCP over control microcosms, and a larger pilot scale study was conducted with the White Rot fungus. Although results of treatability studies indicated increased treatment rates for PCP, the amendments were not considered to be cost effective management options for full scale application. Another commercial compost additive is currently being studied for the potential treatment of highly contaminated soils prior to application to the land treatment units.

Volume Reduction by Soil Screening. Because treatment time is directly related to the volume of soil to be treated, any reduction in soil volume should reduce treatment time assuming that the volume reduction does not affect soil properties that would adversely affect microbial activity. Prior to 1994, contaminated soil was screened through a one inch (2.54 cm) screen before being applied to the land treatment unit. Because the Libby soil contains a significant amount of medium to coarse gravel rock fragments it was concluded that reducing the screen size to 5/8 inch (1.6 cm) would result in a 12 to 14% reduction in contaminated soil volume. The reduction in screen size was considered to be cost effective because estimated cost savings expected through reduced treatment time due to the reduction in volume exceeded the implementation costs of the modified screening process.

Treatment Under Low Oxygen Conditions. Studies conducted at Utah State University (USU) with soil samples from the Libby site in 1995 concluded that degradation of ^{14}C-PCP and ^{14}C-pyrene in microcosms incubated under low soil gas oxygen conditions (2-5%) proceeded at rates that were equal to (pyrene) or significantly higher (PCP) than those of treatments incubated at higher oxygen concentrations (Hurst et al., 1996; Hurst et al., 1997). Considering this information, it was proposed that the rate of lift application could be increased at the Libby site if sufficient oxygen concentrations could be maintained in lower lifts to facilitate continued treatment. Laboratory measurements of gas diffusivity constants and oxygen uptake rates indicated that lower lifts may be supplied with sufficient oxygen (2-21%) through diffusion if this strategy was employed. However, field measurements of soil gas oxygen measurements at various depths in the land treatment units indicated that bioventing technology may be necessary to supply lower lifts with sufficient oxygen.

Soil Mixing. Subsequent studies at USU have included the issue of incorporation of contaminated soil with treated soil to reduce microbial toxicity. Tillage of the lift that is currently being treated results in a zone of incorporation with the underlying treated lift that contains lower contaminant concentrations and an adapted microbial community. To test the effect of an increased zone of incorporation on pyrene degradation, various ratios of treated LTU soil and untreated waste pit soil were mixed and spiked with ^{14}C-pyrene. Results indicated that pyrene degradation rates increase significantly as the ratio of treated to untreated soil increases, but the incorporation of treated and untreated soil only appears to have an affect at mixing ratios of 1:1 or higher. Applied lift depth could be reduced to facilitate this kind of mixing or deep tillage equipment could be employed. Further study is needed to evaluate the cost effectiveness of each approach.

Management of Environmental Factors. The effects of management of environmental conditions including soil moisture and temperature on pyrene degradation have also been recently studied. Both factors were found to have significant effects. Reduction in soil moisture from 85% of field capacity to 40% of field capacity was found to severely inhibit mineralization of ^{14}C-pyrene. This may become an issue at larger soil treatment facilities such as the Libby facility where maintenance of adequate soil moisture during dry periods may be problematic. Use of real time moisture monitoring techniques may be warranted in some cases to prevent fluctuations in soil moisture content.

Increasing temperature to 30 °C also had a significant effect on the rate of ^{14}C-pyrene mineralization, while the rates and extent of pyrene mineralization at 10 and 20 °C were not significantly different from each other. Increasing soil temperature to facilitate increased degradation rates may be difficult. Application of black or clear polyethylene plastic mulch may be used to increase average soil temperatures and decrease the amplitude of the diurnal soil temperature cycle. Further pilot studies are needed to show if this approach may be applicable at sites where contaminant degradation has been shown to occur at low soil oxygen concentrations.

Physical Changes in the Treatment Facility. Finally, the ultimate constraint of treatment time, treatment area, has been modified to facilitate the timely cleanup of the contaminated soil at the Libby site. Initially, the facility consisted of only one, 1-acre land treatment unit. As initial treatment rates made it apparent that this was not sufficient, a second LTU was added in 1991. As treatment continued, and other management options were explored, it was proposed that the soil treatment facility again be expanded to increase the rate of treatment.

The expanded landfarm concept was proposed in 1996 by Champion and leaching and migration studies were used to evaluate the potential impacts on groundwater. Following approval by the U. S. EPA, the expanded landfarm was installed in 1998 increasing the total treatment area of the Libby facility to approximately 14 acres (5.7 ha). Based on the previous treatment history, this expansion should allow for site closure by 2005.

REFERENCES

Hurst, C. J., R. C Sims, J. L. Sims, D. L. Sorensen, J. L. McLean, and S. Huling. 1996. "Polycyclic Aromatic Hydrocarbon Biodegradation as a Function of Oxygen Tension in Contaminated Soil." *J. Haz. Mat. 51*: 193-208.

Hurst, C. J., R. C Sims, J. L. Sims, D. L. Sorensen, J. L. McLean, and S. Huling. 1997. "Soil Gas Oxygen Tension and Pentachlorophenol Biodegradation." *J. Env. Eng. 123*: 364-370.

U.S. EPA. 1996. *Bioremediation Field Performance Evaluation of the Prepared Bed Land Treatment System: Volumes I and II.* EPA/600/R-95/156. U.S. Environmental Protection Agency, Washington, DC.

THE VALIDITY OF EROGOSTEROL-BASED FUNGAL BIOMASS ESTIMATE IN BIOREMEDIATION

Usha Srinivasan[1] and *John A. Glaser* (USEPA, National Risk Management Research Laboratory, Cincinnati, OH 45268
[1]Current Address: Northern Research Institute, Yukon College, Whitehorse, Yukon, Canada

ABSTRACT: The use of lignin-degrading fungi to treat soil contamination has been shown to offer strong potential as a bioremediation technology (Paszczynski, & Crawford,1995). For the proper application of fungal bioremediation, it is important to accurately determine the fungal activity and its distribution throughout a remediation site at different stages of biodegradation. Numerous methods have been explored to measure fungal biomass. Ergosterol, the predominant sterol of most fungi, has been a fairly good indicator for determining fungal biomass especially in soil, without the interference of prokaryotes (Newell, 1992). However, there are certain downfalls in using ergosterol as a biomass indicator (Bermingham et al, 1995). Ergosterol is also common to other soil fungi other than the remediation inoculum and its concentration could vary with the fungal life cycle. Furthermore, it is important to understand any biosynthetic inhibition of ergosterol present in targeted contaminated soils which may limit its utility for the purposes stipulated above (Barrett-Bee & Ryder, 1992). In this study we have attempted to determine the level of ergosterol produced over a period of time by lignin-degrading fungi that have been main bioremediation candidates and also compare their ergosterol levels to other soil fungi that may pose as an interference.

INTRODUCTION

Fungal community size is difficult to determine using environmental samples. Measurement of soil fungal populations has been conducted by several approaches: culture isolation techniques, direct visualization methodologies, or other indirect measurement techniques. Culturing techniques are strongly biased by the choice of nutrient media since these conditions select for components of the soil community that can thrive on the mixture of growth substrates available in the medium.

Indirect features (biochemical and physiological) relating to the cultivation of fungi in contaminated materials to determine activity and amount of active biomass have been investigated. Plate count dilution techniques and assessment of growth in visual estimation on a rank basis were not suitable for dimensional growth patterns of the fungus in soil. Approaches have ranged from the use of mycosteroids, and dye decolorization to PCR detection of ligninase activity. Fluorescein diacetate screening has been used to detect fungal strains inoculated into soil through esterase activity detection. Mycelial growth of *P. chrysosporium* on pine wood chips was assayed by a spectrophotometric fluorescein diacetate method. However, hydrolysis of fluorescein diacetate by *P. chrysosporium* was found to be considerably less than by background resident microflora.

The infestation of cereal grains by fungi has been assayed by monitoring the presence of a mycosteroid, ergosterol (ergosta-5:6,7:8,22:23-trien-3-ol) which has been employed as a quantitative means to determine growth of the fungus. It has been shown to correlate well with the visual estimation technique. Ergosterol analysis has been shown to have a wide range of applications. This information supports the applicability of this technique to the monitoring of fungal biomass throughout a bioremedation application. No definitive correlation of ergosterol expression as a function of growth stage common to all fungal communities, is reported in the literature.

In this study the growth rate of three fungal groups, brown rot, white rot and a few mould fungi were monitored on different media types and the expression of ergosterol by fungal cultures throughout the growth phase and its utility as a means to monitor the active fungal biomass in bioremediation settings.

MATERIALS AND METHODS

Fungal cultures tested. The following fungi were tested - Basidiomycetes : Brown rot fungi - *Postia placenta (MD-506), Sistotrema brinkmannii (ME-632), and* White rot fungi - *Bjerkandera adusta (FP-135160-Sp), Irpex lacteus (Mad-517), Phanerochaete chrysosporium (BKM-F-1767), Phanerochaete sordida (HHB-8922-Sp), Phlebia brevispora (MD-192), Phlebia subserialis (RLG-10693-Sp), Trametes versicolor* (MD-277); Deuteromycetes - *Trichoderma viride (60) and Trichoderma pseudokoningii (64).*

Inoculation procedure, incubation conditions and media preparation. All the fungi were grown on 2% MEA. All the cultures of the treatment plates were incubated at 24 C for growth. The media were all autoclaved at 121 C, 15 psi for 20 min for sterilization. Each treatment was done in triplicate for each fungus.

Preparation of V8J agar, SSE and SSE + BAP or PCP. The fungi were inoculated on 5% V8-J agar (Senior, Epton and Trinci, 1987) and SSE (Soil Solution Equivalent) media (Angle et al., 1991). Only *P.chrysosporium* was inoculated on SSE + bap or pcp plates. After autoclaving the SSE was cooled and 10 ppm of PCP or 350 ppm of BAP was added to the media in flasks before pouring into plates.

Harvest of fungi for ergosterol assay. All the fungi were grown on all three media in triplicate and harvested sacrificially after 2, 4, 8, 12, 16, 20, 24 and 28 days of growth. Harvest procedure - Measured diameter of growth of the fungus, half of the fungal mycelium was for extraction the other half for estimating the biomass.

Ergosterol extraction procedure for V8, SSE and SSE + BAP or PCP. The fungal mat and agar from V8 and SSE media were subjected to total alkali-extractable ergosterol procedure. The SSE with bap or pcp samples were extracted as per the total neutral-extractable ergosterol procedure (Davis and Lamar, 1992).

HPLC analysis. Chromatographic separation over a Vyolde 20 ITP-C18 reverse phase analytical column (0.46 x 25cm) was preceded by a Vydac 2IIECC54T reverse phase C18 guard column. Extracts (20ul) were injected and eluted with MeOH at 45 C at a flow rate of 1ml/min. Column eluant monitored for absorbance at 282 nm. The isolated peak eluting at ca. 8 min retention time was identified as ergosterol. Standard curve for ergosterol was linear for concentration range of. 3 - 100 ug/ml.

RESULTS AND DISCUSSION

In examining the results of this study it is clear that the relationship between "ergosterol" and other factors, such as the genus/species of fungi, nutrients, contaminants and time period of the growth cycle all have a complex relationship. Varied results have been reported for fungal ergosterol analysis where the technique is used to assay the presence of fungi biomass rather than assessing the changes of fungal biomass as a function of some process. Seitz et al. (1979) reported very little variation among grain-colonizing imperfect fungi (*Alternaria and Aspergillus*). However Salmanowicz and Nylund (1988) demonstrated a total range of ± 12% around the mean among four ectomycorrhizal fungi. This emphasizes the importance of studying each of the specific genera of interest carefully.

The average concentration of ergosterol produced in the 28 day life cycle of all the fungal genus was measured in the two media types V8 and SSE medium. It is evident from the results that there is a direct correlation between nutrient level of the growth media and the concentration of ergosterol released by the fungi. Such influences to ergosterol level by nutrient addition may also compromise the sampling results in bioremediation plots where existing nutrients and further amendments made to the soil are unevenly distributed. The range of ergosterol produced between the three fungal groups brown rot , white rot and mold fungi in the two media also seem to show tremendous differences. In the V8 medium, the seven white rot fungi exhibited expressed ergosterol concentration ranges from approx. 1 to 76 ug/g; the two brown rots strains 0 to 35 ug/g, and for the two moulds a larger range of 9 to 100 ug/g. In the SSE medium, they exhibit lower ranges - white rot , 0 to 60 ug/g; brown rot, 0 to 15 ug/g and the moulds 3 to 32 ug/g.

Such differences in ergosterol concentration between the fungal physiological groups under different nutrient circumstances is likely to render problems in monitoring growth rate of a bioremediation agent. In remediation plots even if the soil is initially fumigated to remove presence of other fungi, the disinfection is not full proof. They can colonize the soil as secondary degraders in parts of soil with lower toxicity as bioremediation progresses. Since ergosterol is a universal growth hormone among fungi, the presence of other fungal genera such as *Trichoderma spp.* which may release a higher level of ergosterol as seen in the results above will interfere in estimating the growth rate of the a selected fungal inoculum for bioremediation..

Apart from differences between fungal groups, it is evident from the results that there are marked differences between genera of the same physiological group and even within each genera. For example, in the V8 medium even within the white rot fungi there is a marked difference in ergosterol production, *Phanerochaete sordida* ranges from 9 to 76 ug/g while *Phlebia brevispora,* also a white rot fungi, ranges only from 1 to 16 ug/g. Such differences within white rot fungi is important to note while choosing them as a bioremediation agent, especially if mixed cultures are implemented in the remediation process.

The governing factors for ergosterol production become more complex for observing the concentration of ergosterol produced at different time periods during the fungal growth cycle. Each growth spurt of producing vegetative hyphae or spore structures produce corresponding levels of ergosterol for each component of individual growth cycle of the fungi. Figure 1. shows the concentration of ergosterol produced at different time periods of growth in V8 media in the seven white rot fungi. It is evident from Figure 1. that each of the white rot fungi evaluated, irrespective of belonging to a common genus, produce specifically higher concentration of ergosterol at certain times of their growth period. On the 4th day, *B. adusta, I. lacteus, Ph. subserialis, P. chrysosporium,* fungi from different genus groups, show the highest level of ergosterol concentration than any other time in their 28 day growth cycle. It is evident from this analysis that the time of sampling of a fungal inoculum is an important factor.

The results also show that similarities within the same genus groups, in ergosterol peak times cannot always be expected. In the genus *Phanerochaete*, both the species tested *P. chrysosporium* & *P. sordida*, show a peak in ergosterol production on day 4 and decrease to the same amount by day 8. However the latter species after this steep decrease, gradually increases its ergosterol conc. peaking at an all maximum on day 24. The former fungal species, *P. chrysosporium*, does not follow suit, its ergosterol conc. decreases further after day 8 and actually reaches an all low on day 24 and then shows signs of increase. It is possible that further observation of the same could have shown a similar second peak in ergosterol conc. like the other species. Although some similarity is evident in the ergosterol pattern in the above genus, it is not so with the genus *Phlebia*. These results strongly suggest that each fungal species has a unique growth cycle and subsequently an expressed ergosterol concentration pattern.

The ergosterol pattern in the SSE media for the white rot fungi also exhibited interesting results. *B. adusta* showed an unusually high level of ergosterol on day 12 (61.83 ug/g), falls to a zero and increases again to 41.69 ug/g on day 24. The same species did not exhibit such unusual pattern in the V8 media, which stresses the importance of media influence on ergosterol. The highest level of ergosterol an approx. conc. of 60 ug/g was produced on day 8 in V8 media and on day 12 in the SSE media by *B. adusta*. Therefore, it is crucial to examine life cycles on appropriate nutrient media. The ergosterol conc. among the other white rot fungi also show some unusual pattern. The ergosterol conc. seem to peak and fall more frequently than in V8 media. *P. chrysosporium* & *P. sordida*, both peak on days 4 and 12, but further the former species peaks at day 24 and the latter at day 20 and 28. It is important to note this pattern of frequent rise and fall in ergosterol in this media, as SSE is designed to be closer in nutriency to that of soil, it is possible such variations also exist in soil.

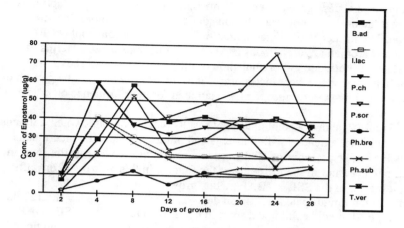

Figure 1. Ergosterol Expression with Time in V8 Media

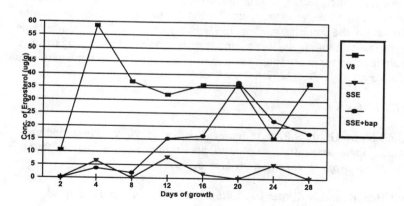

Figure 2. Ergosterol Expression by *P. chrysosporium* in V8, SSE +/- BAP

To add to the complex relationship between,culture nutritional status, fungal genus/species, growth rate and corresponding ergosterol production is an additional factor that is important to consider especially in a bioremediation situation. The influence of contaminants such as pentachlorophenol (PCP) or polynuclear aromatic hydrocarbons (PAH's) must be considered in terms of substrate contribution to culture nutrition or inhibition of ergostrol expression., PCP or benzo-a-pyrene were incorporated into the SSE media and the growth of the white rot fungus *P. chrysosporium* examined for their influence on ergosterol expression.

Figure 2, shows a combination of results that are possible when a fungus is inoculated in soil amended with nutrients and in the presence of contaminants such as BAP. The presence of V8 like media or other rich nutrients the spurt in growth rate is earlier in the life cycle of the fungus, in this case by 4th day after which it tapers. In the SSE +/- BAP media, growth is similar up to 12th day, in both media the fungus peaks a little on day 4 and then day 12. After the 12th day, the SSE + BAP media exhibited a spurt in the growth rate unlike SSE - BAP where peaking occurred on day 20. This response is probably due to the fungal utilization of the BAP as a substrate. Even at the highest level of ergosterol production under these conditions, its concentration is roughly half the concentration reached in the V8 media.

The results of this study show that growth rate of most fungi seem to keep the same pace in V8 and SSE. Although, the density of sporulation was influenced by nutrient content. There is a clear difference in level of ergosterol produced in the media types, and mostly higher in V8 than SSE media. The maximum ergosterol released during the 28 day growth period also seem to vary between the fungal species. Also the ergosterol level peaks and wains at different stages of the growth cycle which seems to be unique to each fungal species, and is influenced by the nutrients of the growth media. Taking all the above observations into consideration, the complex relationship between ergosterol, fungal life cycle, nutrients and contaminants, it is clear that a thorough understanding of such interrelationship for every fungi to be used as an inoculum in remediation studies is necessary.

REFERENCES

Angle, J.S., McGrath, S.P. and Chaney, R.L. 1991. "New culture medium containing ionic concentrations of nutrients similar to concentrations found in the soil solution." *Appl. Environ. Microbiol. 57*: 3674-3676.

Barrett-Bee, K. and N. Ryder 1992. *Biochemical aspects of ergosterol biosynthesis inhibition*, Chapman and Hall, New York, NY

Bermingham, S., L. Maltby, et al. 1995. "A critical assessment of the validity of ergosterol as an indicator of fungal biomass."*Mycol. Res. 99*(4): 479-484

Paszczynski, A; R.L. Crawford,1995. "Potential for bioremediation of xenobiotic compounds by the white-rot fungus." *Biotechnol. Prog. 11*(4): 368-379.

Lamar, R. T., M. W. Davis, et al. 1994. "Treatment of a pentachlorophenol- and creosote-contaminated soil using the lignin-degrading fungus Phanerochaete sordida: A field demonstration." *Soil Biol. Biochem. 26*(12): 1603-1611.

Senior, D.P., Epton, H.A.S. and Trinci, A.P.J. 1987. "An efficient technique for inducing profuse sporulation of *Alternaria brassicae* in culture." *Trans. Bri. Mycol. Soc. 89* (2):244-246.

Seitz, L.M., Sauer, D.B., Burroughs, R., Mohr, H.E. and Hubbard, J.D. 1979. *Phytopath. 69*:1202-1203.

Newell, SY 1992. " Estimating fungal biomass and productivity in decomposing litter," In: *THE FUNGAL COMMUNITY: ITS ORGANIZATION AND ROLE IN THE ECOSYSTEM SECOND EDITION.* Carroll, GC; DT, Wicklow (eds.) MARCEL DEKKER, INC., NEW YORK, NY, pp. 521-561

Salmanowicz, B and Nylund, J.-E. 1988. *Euro. J. Path. 18*: 291-298.

DESIGN AND PERFORMANCE EVALUATION OF VAPOR-PHASE BIOFILTERS

Addanki V. Krishnayya, John G. Agar, and *Tai T. Wong*
(O'Connor Associates Environmental Inc., Calgary, Alberta, Canada)

ABSTRACT: Several past studies have indicated that off-gases resulting from soil and groundwater remedial operations can be treated economically by using biofilters to reduce air pollution. The efficiency of a biofilter depends on several factors such as: vapor retention time, uniformity of vapor flow, composition of the filter, type of microorganisms available for biodegradation, moisture content, and temperature. This paper presents the results of bench scale tests performed on two biofilters at room temperature to evaluate their performance at different gas flow rates and gasoline contaminant concentrations. Based on the test results it was possible to identify some of the key parameters affecting the design and operation of biofilters and to develop rational procedures for the application of biofilters to the treatment of gasoline vapors.

INTRODUCTION

Biofiltration is the process in which an air stream containing volatile organic chemicals (VOCs) is passed through a biologically active medium. The contaminants are transferred from the air phase to the water/biofilm phase, where they are biologically degraded by microorganisms to less harmful substances, usually to carbon dioxide and water (Shareefdeen et al., 1993). Biofiltration has been identified as an economical and efficient method for treating VOCs present in air streams at low concentrations. An excellent review of the various aspects of biofiltration was presented by Wani et al. (1997).

The motivation for research on biofilters at O'Connor Associates Environmental Inc. (OAEI) was mainly due to the simplicity and economy involved in the biofiltration of VOCs at low concentrations resulting from long-term vapor extraction of contaminated soils. The purpose of the study was to identify some of the key parameters affecting the design and operation of biofilters and to develop rational procedures for the application of biofilters to treatment of the VOCs resulting from vapor extraction of gasoline contaminants from soils. The evaluations and conclusions presented here were based on the research data collected during the summer and fall of 1996 and during the years 1997 and 1998.

MATERIALS AND METHODS

Initially, biofiltration experiments were conducted in 1996 by modifying a low profile air stripper made up of a rectangular water storage tank at the bottom, three rectangular stacked trays with perforated bases, and a top cover. The trays were filled with 1 m^3 of biofilter material. Air mixed with gasoline vapors was bubbled through the water in the storage tank. The contaminated moist air entered at the base of the bottom tray and passed upward through the three trays. Experiments carried out at various flow rates and vapor concentrations indicated that the contaminant removal efficiency of the biofilter was significantly lowered due to non-uniform airflow caused by flow channels and short-circuiting. The removal efficiency of the biofilter during operation is calculated using

$$E_o = 100(C_i - C_o)/C_i \tag{1}$$

where E_o is the biofilter efficiency during operation in %, C_i and C_o are the contaminant concentrations of inflow and outflow gas in ppm respectively.

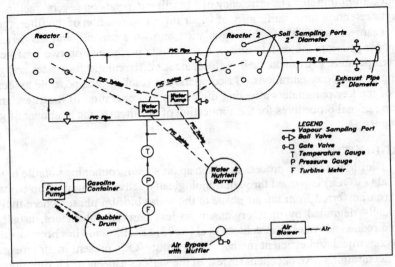

Figure 1. Schematic of Biofilter Experiment Setup

Because of the low efficiencies obtained in the earlier experiments, the experimental set up was modified to that shown in Figure 1. Two biofilter reactors were constructed using plastic drums; each was 0.79 m in diameter and 1.12 m in height. Tables 1 and 2 summarize respectively the composition and the properties of the biofilter material used in the experiments.   The spent

Table 1. Composition and Cost of Biofilter Material

Material	% by Volume
Peat moss	18.0
Vermiculite	7.5
Lime stone	2.0
Pine bark	20.0
Organic compost	30.0
Spent activated carbon	15.0
Bacteria triggering mechanism (BTM)	2.0 L/m³ of filter
Humic acid	0.5 L/m³ of filter
KNO₃	90 g/m³ of filter

activated carbon was previously exposed to gasoline vapors. The unit cost of the biofilter material was about $100 (Canadian) per cubic meter. Biofilter material was placed in each reactor to a thickness of 0.76 m and was supported by a deck built at a height of 0.15 m above the base. A water sprinkler was installed at the top of each reactor to add water and nutrients when necessary.

Table 2. Properties of Biofilters in Reactors

Final volume per reactor	0.37 m³ (13.1 ft³)
Porosity	0.63
Gravimetric water content	133.2%
Volumetric water content	45.2%
Air porosity	0.173
Air pore volume per reactor	0.064 m³ (2.265 ft³)

Six ports were installed on the top of each reactor for measuring vapour concentration and sampling. As shown in Figure 1, the biofilter reactors could be connected either in series or in parallel configuration. A 1-h.p. blower was used to supply air to the biofilters. Air was bubbled through water in a plastic drum to which fresh unleaded gasoline was added using a chemical feed pump. The supply of humidified gasoline vapors to the biofilters prevented loss of moisture in the filters and helped in maintaining high relative humidities (95 to 98%) in the exhaust vapors. Typically, each test would last between 60 to 90 min.

Vapor inflow and outflow rates were determined by measuring velocities using an anemometer and by metering airflow through a turbine meter. Vapor concentrations were measured using a gas detection meter (GasTech Model 1238ME). Also vapor bag samples were collected and analysed for BTEX using a HNU Model 311 portable gas chromatography (GC) system. BTEX is the group of aromatic hydrocarbons including benzene, toluene, ethylbenzene and xylene. All experiments were conducted at ambient air temperatures ranging between 20°C and 24°C. For the biofilter material, total and hydrocarbon-degrading micro-organisms counts, pH, organic carbon content, and nitrate concentration were also measured. An increase in plate count of total microorganisms from 3.2×10^8 to 1.8×10^9 CFU/g was observed to occur in about two months. The hydrocarbon-degrading micro-organisms count was 1.0×10^8 CFU/g. The pH of the filter medium varied between 6.5 and 7.5. The organic carbon content varied between 23 and 33%. Nitrate concentrations were between 100 and 170 ppm.

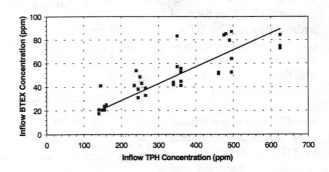

Figure 2. Inflow BTEX Versus TPH Concentrations

The relationship between the inflow gasoline vapor concentrations as measured by the gas detector and the BTEX

concentrations as deter-
mined by GC analyses is
presented in Figure 2. For
the experiments reported in
this paper, unleaded gasoline was used. The aver-
aged BTEX concentrations
were between 12% and
17% of the total petroleum
hydrocarbon (TPH) concen-
tration. The percentage of
BTEX concentration was
slightly higher at low inflow concentrations.

Figure 3 presents
the TPH removal efficiency
versus vapor retention time
when the two reactors were
connected in series. It is
apparent that the removal
efficiency is increased with
longer retention time and
lower inflow vapor concentration. Figure 4 shows the
total BTEX removal effi-
ciency versus vapor retention time when the reactors
were connected in series.
Significantly higher removal efficiencies were achieved for BTEX than for total
hydrocarbons. The BTEX
removal efficiencies also
increased with longer retention time and lower inflow
concentrations.

Figure 5 presents
the efficiency in the removal of individual components
of BTEX, namely benzene,
toluene, ethylbenzene and
xylenes versus the inflow
total BTEX concentration.
The removal efficiencies
for benzene were slightly

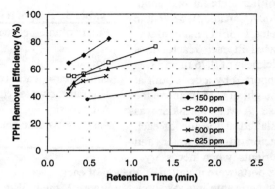

**Figure 3. TPH Removal Efficiency Versus
Retention Time, Reactors in Series**

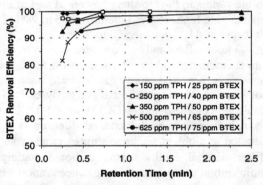

**Figure 4. BTEX Removal Efficiency Versus
Retention Time**

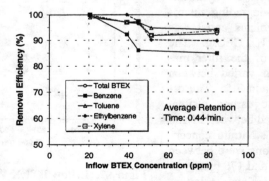

**Figure 5. Removal Efficiency Versus Inflow
BTEX Concentration**

lower than the other aromatic hydrocarbons.

The performance of the individual biofilters (Reactors 1 and 2) was also studied by connecting the reactors in parallel. Trends in the variation of the removal efficiencies with retention time and vapor concentration were similar to those described for the experiments conducted in series.

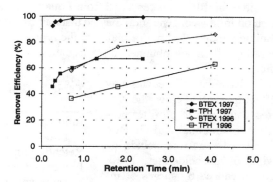

Figure 6 compares the removal efficiencies of experiments conducted in 1996 to those from experiments conducted in 1997/98. In 1996, the reactor used was made up of stacked rectangular trays, while cylindrical

Figure 6. Effect of Non-Uniformity in Vapor Flow on Biofilter Efficiency

drums were used in 1997/98. The efficiencies were compared for experiments conducted at an inflow total hydrocarbon concentration of 350 ppm and an inflow BTEX concentration of 50 ppm. A significant improvement in efficiencies was noted in the 1997/98 experiments because of the improved uniformity in vapor flow conditions achieved in the cylindrical drum type reactors. Figure 6 clearly demonstrates the importance of the uniformity of vapor flow through biofilters in achieving higher efficiencies.

BIOFILTER DESIGN

The required or targeted efficiency of a biofilter in the removal of a given contaminant by degradation may be expressed as:

$$E_t = 100\left[1 - \frac{694.4M}{mC_iQ}\right] \tag{2}$$

where E_t is the targeted biofilter efficiency in %, M is the permissible mass output of the contaminant into atmosphere in kg/day, m is the conversion factor from ppm to mg/m^3 (for benzene = 3.24 at 20°C), C_i is the contaminant concentration of inflow gas in ppm, and Q is the gas flow rate through the biofilter in m^3/min.

The required biofilter volume is calculated from

$$v = Qt/n_a \tag{3}$$

where v is the volume of biofilter, t is the retention time and n_a is the air porosity of biofilter.

The targeted biofilter efficiency is obtained from Equation 2. A suitable retention time for the biofilter to achieve the required efficiency is chosen and the volume of biofilter is determined from Equation 3. Selection of the retention time based on inflow concentration may be made from the data presented in this paper for gasoline vapors. The performance of the biofilter is considered satisfactory if the operating efficiency obtained from Equation 3 is maintained above the targeted efficiency.

Removal efficiencies may fall temporarily below the targeted value due to biofilter properties, operational and climatic factors. These factors include the drying of the bio-filter, break through caused by surges of inflow rate and/or inflow concentration, development of non-uniform flow caused by air channels and short-circuiting in the biofilter, aging of the biofilter due to biomass development, and reduction in biofilter temperature due to cold weather. Table 3 presents the desirable range for the various parameters necessary for the successful operation of biofilters for treating gasoline vapors. This table is compiled using data collected from the tests discussed in this paper as well as from the literature. In addition to these parameters, proper maintenance of biofilters is highly important to ensure their successful operation at desired efficiencies.

Table 3. Biofilter Design and Operation Parameters

Inflow vapor concentration	< 500 ppm
Retention time	0.5 – 2.0 min.
Empty bed contact time	2.5 - 10.0 min.
Air pore volume of biofilter	15 – 30 %
Gravimetric moisture content (by dry weight)	110 – 140 %
Porosity of filter	60 – 70 %
Organic carbon content	20 – 30 %
Nutrients in water phase	100 – 200 ppm
Oxygen in outflow vapors	> 20 %
pH value	6.5 – 7.5
Temperature of filter	15 – 30 °C
Relative humidity of inflow vapor	95 – 98 %

REFERENCES:

Shareefdeen, Z., B.C. Baltzis, Y.S. Oh, and R. Bartha . 1993. "Biofiltration of Methanol Vapor." *Biotechnology and Bioengineering*, 41(5): 512-524.

Wani, A.H., R.M.R. Branion, and A.K. Lau. 1997. "Biofiltration: A Promising and Cost-Effective Control Technology for Odors, VOCs and Air Toxics." *J. Environ. Sci. Health*, A32(7): 2027-2055.

BIOMASS DISTRIBUTION IN A VAPOR PHASE BIOREACTOR FOR TOLUENE REMOVAL

JiHyeon Song & Dr. Kerry A. Kinney

Department of Civil Engineering

The University of Texas at Austin

ABSTRACT: The objectives of this research were to determine the spatial and temporal changes in biomass distribution in a vapor phase bioreactor, and to examine the effects of nutrient availability on bioreactor performance. A lab-scale bioreactor containing pelletized ceramic media was operated at a toluene loading rate of 45.8 g/m^3hr. To supply moisture and nutrient to the biofilm, a small amount of modified hydrocarbon minimum media (HCMM) was supplied to the biofilter by means of a nutrient laden aerosol. When the nutrient supply was limiting, the overall bioreactor removal efficiency declined to approximately 80% and the toluene removal profile was a linear function of bed depth. However, after adequate nutrients were supplied to the biofilm, an overall removal efficiency of >99% was achieved, and greater than 90% of the toluene was removed in the front half of bed corresponding to an elimination capacity of 80 g/m^3hr. The toluene degrading capacity of the front half of the bioreactor slowly decreased after a short period of pseudo-steady state due to excess biomass build-up, a decrease in moisture content and inactivation of the biomass. Biomass distributions determined by volatile solids (VSs) and total plate counting indicated that excess biomass in the inlet bioreactor section was responsible for the decline in bioreactor performance.

INTRODUCTION

Biofiltration of volatile organic compounds (VOCs) has received more attention recently, as vapor phase bioreactors are being recognized as a potentially cost-effective alternative to traditional air pollution control methods. However, in many experimental investigations, bioreactor performance is characterized by determining only the overall pollutant removal efficiency of the system. This "black box" approach may result in erroneous interpretation of bioreactor performance and neglects changes in biomass density and reaction kinetics that occur in biofilters.

Since vapor phase bioreactors rely on microorganisms growing on the surface of support media, determining the biomass distribution is essential to characterize the biofilm. For long-term operation of bioreactors, spatial and temporal changes in microbial parameters such as biomass density can have a substantial effect on bioreactor performance. The microbial parameters as well as the pollutant degrading efficiency should, therefore, be carefully monitored to maintain stable operation and to control the bioreactor. However, most studies of vapor phase biofiltration neglect these parameters, and the biomass density is often assumed to be constant with time and space.

Biomass density in bioreactors treating easily degradable compounds is greatest near the gas stream inlet where the carbon supply is abundant (Ergas et al., 1994). Several recent studies have focused on the distribution of biomass in vapor phase bioreactors (Smith, et al., 1996; Kinney, 1996) because this uneven distribution can result in biomass clogging and a high pressure drop across the bioreactors. Nevertheless, little work have been done to quantitatively correlate biomass distribution to bioreactor performance.

The objectives of this study were (1) to understand the microbial response of a biofilm reactor subjected to a high loading rate of toluene; (2) to determine the effect of system performance on biomass accumulation; and (3) to examine nutrient delivery requirements necessary for stable operation of the bioreactor.

MATERIAL AND METHODS

Bioreactor Configuration and Operation. A lab-scale bioreactor consisting of a stainless steel column with a diameter of 16.2 cm and an overall height of 100 cm was operated for 96 days. The reactor column was divided into four sections and each section contained pelletized packing media (Celite® R-635, Lompoc, CA) to a depth of 25 cm. An air stream contaminated with 200 ppm$_v$ toluene was supplied at a flow rate of 19.8 L/min corresponding to an overall toluene loading rate of 45.8 g/m^{3}hr and an empty bed residence time of 1 minute. The experimental setup is shown in Figure 1.

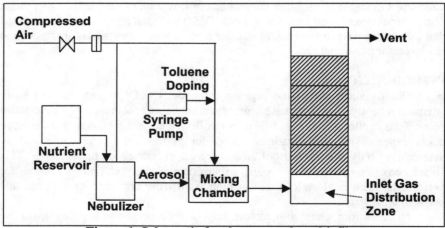

Figure 1. Schematic for the vapor phase biofilter

Nutrients and moisture were supplied using a nebulizer (Heart®, Tucson, AZ) to generate a fine aerosol. The fine aerosol contained a nutrient-buffer solution, which was mixed with the inlet air stream. The nutrient solution initially used was a hydrocarbon minimal media (HCMM) slightly modified from the composition described elsewhere (Ridgway, et al., 1990), and consisted of a phosphate buffer, two nitrogen sources (ammonia and nitrate) and trace metals.

To meet the nitrogen requirements of the bioreactor and to provide adequate buffering capacity, the phosphate, ammonia and nitrate concentrations were modified four times over the course of the experiments, while the trace metal concentrations were held constant.

Sampling and Analysis. Each section of the bioreactor had one gas sampling port and two packing media sampling ports. To measure the toluene profile along the column, gas samples were collected daily with a gas-tight syringe, and analyzed using a gas chromatograph (Hewlett Packard 6890) equipped with a flame ionization detector (GC/FID). Packing media samples were periodically collected by withdrawing several packing pellets through the media sampling ports, and analyzed for volatile solids (VSs) and total heterotrophic plate counts. The VSs measurements were determined by weight loss following ignition of the samples at 550 °C, in accordance with Standard Methods (1989). The weight loss upon combustion of control samples of clean packing pellets was negligible. The volatile solids measurements provided a rough estimate of the amount of organic matter (i.e., biomass) in the bioreactor samples. As another indicator of biomass accumulation, colony forming units (CFUs) on R2A agar media (Difco, Detroit, MI) plates were used to determine total heterotrophic cell numbers.

RESULTS AND DISCUSSION

Moisture and Nutrient Supply. A unique feature of the experimental setup was that the moisture and nutrients required for microbial growth and biodegradation were supplied to the bioreactor by an aerosol feed. Since a small amount of HCMM solution (400 mL/day) was delivered directly to the packing media, very little leachate (70-120 mL/day) was produced, even though a sufficient amount of water was supplied to the bed to maintain an adequate moisture content. The aerosol nutrient feed effectively delivered nutrients to the bioreactor while minimizing free liquid holdup that can lead to mass transfer limitations and high pressure drops (Auria et al., 1998).

Bioreactor Performance. After bioreactor start-up, the overall toluene removal efficiency was approximately 80%, but the toluene removal profile was found to be a linear function of bed depth (see Figure 2, day 15). Since the packed biofilter is a plug flow reactor, the linear profile was indicative of the zero order reaction kinetics within the biofilm.

Nitrogen limitation was suspected as the cause of the relatively low toluene removal and zero order kinetics. Since nitrogen is one of the major cell constituents (around 8-13% of dry cell weight), it may be a potential growth limiting nutrient if an adequate amount is not available for the microorganisms. Nitrogen addition has been shown to increase pollutant removal efficiency and restore lost elimination capacity in a compost biofilter (Morgenroth et al, 1996).

To investigate this possibility, the nitrate concentration in the aerosol solution was tripled on day 24 to meet the nitrogen requirement of the biofilm.

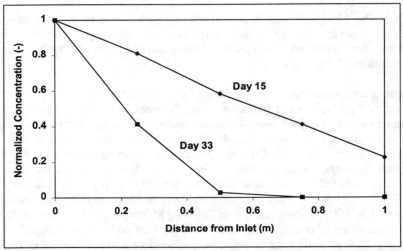

Figure 2. Toluene profiles across the column on day 15 and 33, respectively.

After increasing the nitrate concentration, a first order exponential decay of toluene was observed along the column and an overall removal efficiency of >99% was achieved (see Figure 2, day 33). This concentrated nutrient solution was used for the remainder of the experiment. During the pseudo steady state period (days 25 to 40), greater than 90% of the inlet toluene was removed in the front half of bed corresponding to an elimination capacity of 80 g/m^3hr. The final 10% of toluene was removed in the remaining portion of the bed. These results indicate that nutrient availability is an essential factor for bioreactor start-up and stable operation.

After a short period of pseudo-steady state operation (about 15 days), the toluene degrading capacity in the first half slowly decreased. As shown in Figure 3, the front two sections had lost most of their toluene degrading capacity by day 65, and only 20% of the inlet toluene was degraded in the front half of the bioreactor. The remaining 80% of the toluene was removed in the second half of bed corresponding to an overall removal efficiency of >99%. Eventually, the second half of bed also lost its toluene degrading capacity, as the inactivation of the filter bed progressed from the inlet section to the end of the bioreactor. On day 87, the overall removal efficiency was only 30%. The reason for the inactivation is not clear; however, excess biomass build-up and a decrease in moisture content due to microbial heat production in the inlet section likely contributed. Further research on this inactivation process is currently ongoing.

It is interesting to note that the toluene removal profiles along the column on day 33 and day 65 were absolutely different, even though the overall removal efficiencies were both >99%. On day 33, the actual elimination capacity was very high, indicating that the inlet toluene loading could be effectively handled in the first half of the biofilter and in the back half, if necessary. However, by day 65, the entire biofilter column was required to achieve the overall toluene elimination

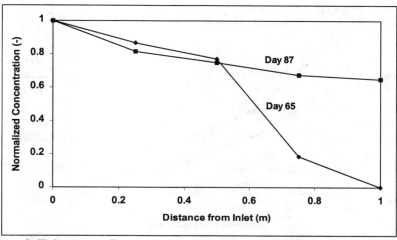

Figure 3. Toluene profiles across the column on day 65 and day 87, showing the loss of toluene degrading activity from the inlet.

capacity of 45.8 g/m³·hr. Thus, the overall removal efficiency provides limited information that could make it difficult to predict bioreactor performance properly, whereas the pollutant and biomass profile along the column provides more insight into microbial response and reaction kinetics.

Biomass Accumulation. As shown in Figure 4, biomass accumulation in the biofilter increased gradually with time. During the nitrogen limitation phase (day 1 to day 20), the VS increase was restricted but evenly distributed, because the toluene degradation took place throughout the entire column. After adequate nitrogen was supplied to the biofilm, biomass accumulated mostly in the inlet section where carbon sources were most abundant. The VS profiles indicate that almost no biomass accumulated in the outlet section, and very little in the third section between days 20 and 60, while the biomass in the inlet section increased 10 fold during the same period. The biomass distributions determined by plate counts on R2A agar media followed the same trend as the VS results. This uneven distribution of biomass caused excess biomass build-up and a high pressure drop across the inlet section (i.e., greater than 10 inches of water throughout the bioreactor), resulting a decrease in toluene degradation. After the front half of bed lost toluene degrading activity, biomass accumulated in the back half of bed which was actively degrading toluene between days 60 and 96.

These results indicate that in order to maintain stable performance it is very important to monitor and control biomass distributions within the bioreactors. Simply monitoring the exit gas phase pollutant concentration does not provide enough information to predict and prevent bioreactor failure.

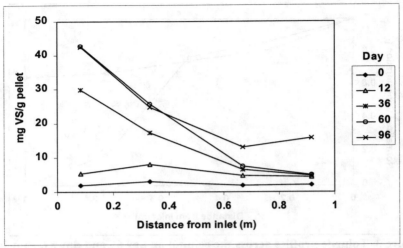

Figure 4. Biomass profiles across the bioreactor with time.

REFERENCES

APHA, AWWA and WPCF. 1989. *Standard Methods for the Examination of Water and Wastewater*, 17th ed., New York.

Auria, R. et al. 1998. "Influence of Water Content on Degradation Rates of Ethanol in Biofiltration," *J. Air and Waste Management Association*, 48: 65-70.

Ergas, S. J. et al. 1994. "Spatial Distribution of Microbial Populations in Biofilters," In *Proceedings of the 87th Annual Meeting of the Air and Waste Management Association*, Cincinnati, Oh.

Kinney, K. A. 1996. *Spatial Distribution of Biomass and Activity in a Directionally-Switching Biofilter*, Ph.D. Dissertation, University of California, Davis.

Morgenroth, E. et al. 1996. "Nutrient Limitation in a Compost Biofilter degrading Hexane," *J. Air and Waste Management Association*, 46: 300-308.

Ridgway, H. F. et al. 1990. "Identification and Catabolic Activity of Well Derived Gasoline Degrading Bacteria from a Contaminated Aquifer," *Appl. & Environ. Microb.*, 56(11): 3565-3575

Smith, F. L. et al. 1996. "Development of Two Biomass Control Strategies of Highly Efficient Biofilters with High Toluene Loading," *Environ. Sci. Technol.*, 30: 1744-1751.

REMOVAL OF TOLUENE IN A FUNGAL VAPOR-PHASE BIOREACTOR

J. R. Woertz (University of Texas, Austin, TX)
K. A. Kinney (University of Texas, Austin, TX)
N. D. P. McIntosh (University of Texas, Austin, TX)
P. J. Szaniszlo (University of Texas, Austin, TX)

ABSTRACT: Toluene is a volatile organic compound (VOC) that is listed as a hazardous air pollutant (HAP). As stricter regulations on VOCs and HAPs have been promulgated, the demand for more cost effective and efficient abatement-technologies has increased. One promising new technology for removing VOCs from polluted gas streams is biofiltration, a process in which contaminated air is passed through a biologically active bed. Although bioreactors that contain bacteria have been successfully employed to treat VOCs in waste gas streams, bacterial systems are generally most applicable for removing low concentrations of VOCs. In this study, a fungal vapor-phase bioreactor, operated under aerobic conditions and a 39-second residence time, was used to treat a gas stream contaminated with varying concentrations of toluene. The fungal bioreactor achieved a maximum toluene elimination capacity of 285 g/m^3-hr, which is three to seven times greater than toluene elimination capacities typically reported for bacterial systems. In addition, harsh operating conditions such as low moisture content and low pH did not adversely affect the performance of the fungal bioreactor. These results indicate that fungal bioreactors may be an effective alternative to traditional abatement technologies for treating high concentrations of VOCs in waste gas streams.

INTRODUCTION

Biofiltration, a process in which contaminated air is passed through a biologically active bed, has long been employed for the purpose of removing odorous compounds at wastewater treatment plants. In the last two decades, however, more emphasis has been placed on using biofiltration for other industrial processes such as treating VOCs in off-gas streams from manufacturing plants (Leson and Winer, 1991). Previous studies have shown that bacterial biofilms in vapor-phase bioreactors are capable of degrading low concentrations of VOCs, including toluene (Seed and Corsi, 1994; Severin et al., 1993). More recently, studies have demonstrated that vapor-phase bioreactors containing fungal biofilms are capable of removing higher concentrations of VOCs from off-gas streams (Cox et al., 1993; Van Groenestijn and Hesselink, 1994); however, only a few VOCs have been investigated as carbon sources, and little is known about the ability of fungal biofilms to degrade toluene.

The objective of this paper is to investigate the feasibility of using fungal vapor-phase bioreactors as viable alternatives to traditional treatment processes for VOCs. The effect of toluene loading, as well as biofilm pH and moisture content, on the bioreactor's toluene removal efficiency is discussed. Furthermore, the results of the study are compared with results obtained in previous studies conducted with bacterial systems. Finally, results from bottle studies are presented, which confirm that the fungi isolated from the bioreactor were responsible for using toluene as their sole carbon and energy source.

MATERIALS AND METHODS

Bioreactor Design. The 14.4-L bench-scale bioreactor (15.5 cm diameter, 75 cm packed-bed depth) used in this study was constructed from stainless steel and was packed with inert, porous silicate pellets (Celite R-635). The bioreactor was supplied with a humidified, synthetic waste gas stream containing 50 to 900 ppm$_v$ toluene. The waste stream was produced by mixing two separate air streams. The first air stream was continuously supplied toluene via a syringe pump (kd Scientific, Model KDS200). The second air stream was fed into a Heart nebulizer (VORTRAN Medical Technology, Inc.). Using approximately 500 mL per day of nutrient media (as described in Table 1), the nebulizer created a fine, nutrient-laden aerosol, which humidified the air stream. The humidified and toluene-laden air streams were then fed into a stainless steel mixing chamber before being passed through the bioreactor. The total flow to the bioreactor was maintained at 22.2 L/min, resulting in an empty bed contact time (EBCT) of 39 seconds. The bioreactor was operated in a directionally-switching mode, and the position of the inlet feed was changed from the top to the bottom of the bioreactor every 7 days. The bioreactor was kept in a constant temperature room maintained at 21°C. A schematic of the system is shown in Figure 1.

TABLE 1. Nutrient media provided to the fungal bioreactor.

Major Nutrients	Concentration (g/L)	Trace Metals	Concentration (mg/L)
KH_2PO_4	4.0	$CuCl_2 \cdot 2H_2O$	0.17
K_2HPO_4	6.0	$CoCl_2 \cdot 6H_2O$	0.24
$(NH_4)_2SO_4$	5.0	$ZnSO_4 \cdot 7H_2O$	0.58
$MgSO_4 \cdot 7H_2O$	0.25	$MnSO_4 \cdot H_2O$	1.01
$CaCl_2 \cdot 2H_2O$	0.02	$Na_2MoO_4 \cdot 2H_2O$	0.24
		$NiCl_2 \cdot 6H_2O$	0.10
		$FeSO_4 \cdot 7H_2O$	1.36

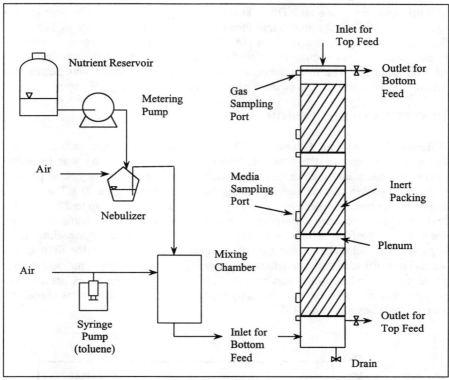

FIGURE 1. Bench Scale Fungal Vapor-Phase Bioreactor System.

Inoculation. The bioreactor was inoculated with *Acremonium* sp. and *Rhinocladiella* sp. growing on Celite packing material. The inoculate was obtained from a previous study at the University of Texas (Woertz, 1998).

Bottle Studies. Pure cultures of toluene-degrading fungi were developed by isolating fungal species from the bioreactor onto R2A plates (Difco, Inc.). The isolates were aseptically transferred into 160-mL serum bottles filled with 50 mL of nutrient media (Table 1) and capped with gray butyl rubber septa and aluminum crimp caps. Six (6.0) µL of toluene were aseptically injected into each bottle using a 100-µL syringe (SGE, Inc.) with a sterilized needle. The headspace of each culture was analyzed for the presence of toluene every two days as described below. The bottles were stored on a shaker table at 21°C.

Analytical Methods. The amount of toluene present in the headspace of the bottles and in the synthetic gas stream was determined by injecting 0.5-mL gas samples onto a Hewlett Packard Model 6890 Gas Chromatograph (GC) equipped with a flame-ionization detector (FID) and a HP-5 capillary column. Ultra high purity (UHP) helium was used as the carrier gas at a flow rate of 16 mL/min. The FID was supplied with 28.4 mL/min UHP helium, 50 mL/min UHP hydrogen and

450 mL/min zero grade air. The FID was maintained at 250°C. The column was operated at 50°C for 2 minutes and then increased to 140°C at a rate of 12°C/min. The column was maintained at 140°C for 3 minutes to remove residual contaminants from the column. Using this method, toluene had a retention time of 3.68 minutes. The moisture content of the packing material was determined gravimetrically, and the pH of the biofilm was measured using pH paper.

RESULTS AND DISCUSSION

Toluene Elimination Capacity.   To determine the maximum elimination capacity of toluene in the fungal bioreactor, the toluene loading was increased every four hours and the elimination capacity (mass of toluene degraded per unit volume of packing media per unit time) was measured (Figure 2). The toluene elimination capacity was linear with respect to toluene loading up to 270 g/m^3-hr, and a maximum toluene elimination capacity of 285 g/m^3-hr was achieved. From the removal profiles shown in Figure 3, it is evident that a higher elimination capacity was achieved in the top 25-cm section of the bioreactor than in the second or third sections. Nonetheless, the maximum toluene elimination capacity achieved by the bioreactor was approximately three to seven times greater than toluene elimination capacities typically reported for bacterial systems (Seed and Corsi, 1994; Severin et al., 1993).

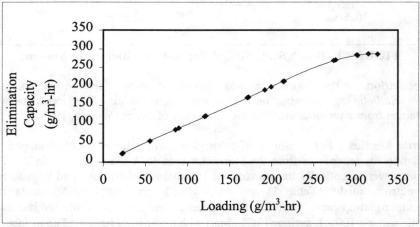

FIGURE 2. Toluene Elimination Capacity at an EBCT of 39 seconds.

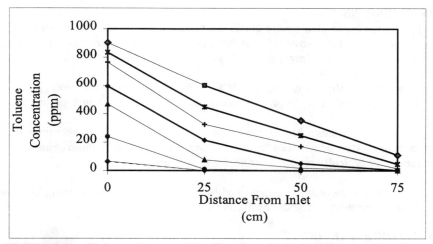

FIGURE 3. **Toluene Removal Profiles for a Range of Toluene Loadings.**

Adaptation to Harsh Environmental Conditions. Two tests were performed to determine how the fungal biofilm would react to harsh environmental conditions, which are often encountered when operating a bioreactor in field applications.

The first test consisted of removing the nebulizer from the process stream, thus creating an arid environment. The biofilm quickly began to dry out, and the moisture content of the biofilm dropped below 20% (wet basis). Although the moisture content remained below 20% for more than 3 weeks, no loss in elimination capacity was observed, and the bioreactor maintained greater than 99% removal of 40 g toluene/m^3-hr.

During the second test, the potassium phosphate buffer was removed from the nutrient media in order to determine how acidic conditions affected the fungal biofilm. The pH of the biofilm fell as low as 3; however, no detrimental effects were observed with respect to toluene removal during the four-week test.

Bottle Studies. In order to determine if both fungi isolated in the bioreactor had the ability to use toluene as their sole carbon and energy source, bottle studies using pure cultures isolated from the bioreactor were performed. It was observed that the *Rhinocladiella* sp. and *Acremonium* sp. were both able to utilize toluene as their sole carbon and energy source. The toluene degradation also resulted in an increase in turbidity in the *Rhinocladiella* sp. and *Acremonium* sp. cultures. The *Acremonium* sp., however, had a much longer lag time than did *Rhinocladiella* sp. The long lag time may have been due to differences in the amount of culture that was transferred from the plates to the bottles. In contrast, negligible losses of toluene and no turbidity increases were observed in the autoclaved control cultures. Additional studies are needed to determine degradation rates and yield coefficients for each culture.

CONCLUSIONS

Numerous experiments were conducted to determine the feasibility of using a fungal bioreactor to remove toluene from contaminated air streams. The major conclusions that can be drawn from this work are given below.

1. Fungi are capable of utilizing VOCs as their sole carbon and energy source and can be used in vapor-phase bioreactors to treat contaminated gas streams.

2. Fungal vapor-phase bioreactors have the potential to achieve elimination capacities three to seven times greater than the maximum elimination capacities typically reported for bacterial systems.

3. Fungal bioreactors are resistant to adverse operating conditions, including low moisture content and acidic biofilms.

ACKNOWLEDGEMENTS

Ms. Woertz would like to thank the National Science Foundation for her funding while conducting her research.

REFERENCES

Cox, H. H. J., J. H. M. Houtman, H. J. Doddema, and W. Harder. 1993. "Enrichment of Fungi and Degradation of Styrene in Biofilters." *Biotechnology Letters*. 15(7): 737-742.

Leson, G. and A. M. Winer. 1991. "Biofiltration: An Innovative Air Pollution Control Technology for VOC Emissions." *Journal of the Air & Waste Management Association*. 41(8): 1045-1054.

Seed, L. P. and R. L. Corsi. 1994. "Biofiltration of BTEX-Contaminated Gas Streams: Laboratory Studies." In Air & Waste Management Association 88th Annual Meeting & Exhibition. Cincinnati, Ohio.

Severin, B. F., J. Shi, and T. Hayes. 1993. "Destruction of Gas Industry VOCs in a Biofilter." In IGT Sixth International Symposium on Gas, Oil, and Environmental Technology. Colorado Springs, Colorado.

van Groenestijn, J. W. and Paul G. M. Hesselink. 1994. "Biotechniques for Air Pollution Control." *Biodegradation*. 4: 283-301.

Woertz, J. R. 1998. Removal of Toluene and Nitric Oxide in a Fungal Vapor Phase Bioreactor. M.S. Thesis, University of Texas at Austin, Austin, Texas.

GAS PHASE BIOTREATMENT OF MTBE

Nathalie. Y. Fortin and *Marc A. Deshusses*
(University of California, Riverside, CA)

ABSTRACT: An aerobic microbial consortium able to use methyl tert-butyl ether (MTBE) as sole carbon and energy source was enriched in two waste air biotrickling filters. After the acclimation phase, the biotrickling filters were able to degrade up to 50 g of MTBE per cubic meter of reactor per hour, a value comparable to other gasoline constituents. The MTBE degrading biotrickling filters were characterized by their almost full conversion (~97%) of MTBE to carbon dioxide, the absence of any degradation by-products in either the gas or the liquid phase, their very high specific degradation activity per amount of biomass, and their low rate of biomass accumulation - a clear advantage for a future industrial application. Overall, the results presented herein demonstrate that although MTBE is still often considered as a recalcitrant compound, it can be effectively biodegraded under carefully controlled environmental conditions.

INTRODUCTION

Depending on the season, reformulated gasoline can contain up to 15% methyl tert-butyl ether (MTBE) by volume. Over the years, the production of MTBE has grown rapidly. At the same time, an increasing number of contaminations of either shallow or deep groundwater with MTBE have been reported (Squillace et al., 1996; Happel et al., 1998) raising high concerns as to the real environmental impact of MTBE. While MTBE target levels for drinking and surface water are being developed, cost effective remediation techniques are needed to treat MTBE contaminations. In many remediation cases (e.g., air sparging, SVE, air striping, or wastewater treatment operations) large air streams contaminated with MTBE are generated that require further treatment. An emerging control technology for this is biotreatment (Devinny et al., 1999). The most promising bioreactors for air pollution control are biofilters and biotrickling filters (Deshusses, 1997; Cox and Deshusses, 1998; Devinny et al., 1999). In biotrickling filters, the contaminated air is passed through a packed bed on which pollutant degrading organisms are immobilized. An aqueous solution is trickled over the bed to provide the necessary moisture and mineral nutrients. Biotrickling filters often exhibit higher performance than biofilters. This is because they allow a better control of environmental conditions and because they rely on growing organisms, rather than on essentially resting organisms as in the case of biofilters. For these reasons, biotrickling filters offer promise for the treatment of MTBE. Hence, the motivation for the present study was to develop an efficient and sustainable process for MTBE vapor control.

MATERIALS AND METHODS

Biotrickling Filter Setup and Operating Conditions. Two parallel biotrickling filters served both for the enrichment of the MTBE degrading consortium and to investigate the performance of MTBE removal from synthetic waste gas. The reactors consisted of a 1.5 m high clear PVC pipe (ID = 0.153 m), filled with 0.5 m of packing material. Reactor 1 was filled with 8.81 kg of wet lava rock (1-3 cm diameter) and reactor 2 was filled with 0.94 kg of 2.5 cm polypropylene Pall rings (Flexirings, Koch Engineering, Wichita, KS). Synthetic MTBE contaminated waste air gas was introduced at the top of each columns. An aqueous mineral medium liquid phase was continuously recirculated over the packed beds at a volumetric flow rate of 0.15 $m^3 h^{-1}$. Fresh mineral medium (in g L^{-1}: K_2HPO_4: 2, KH_2PO_4: 1, NH_4Cl: 0.75, $MgSO_4$: 0.5, $CaCl_2$: 0.018 and 1 mL L^{-1} trace element solution was continuously fed at an average flow rate of 50 mL h^{-1}. Throughout the experiments, the MTBE inlet concentration was maintained between 0.65 and 0.85 g m^{-3} (183-239 ppm_v).

Analyses. MTBE and TBA in liquid samples was analyzed by gas chromatography with a flame ionization detector (FID). Formaldehyde was analyzed with the chromotropic acid method. Grab samples from either inlet or outlet gas streams were simultaneous analyzed for CO_2 and MTBE or other volatiles by GC with a thermal conductivity and a FID detector. To determine the amount of wet biomass in the reactor, the recycle medium was allowed to drain, the reactor was then weighed. The original inoculum for the reactor was groundwater and aquifer material from two long term MTBE contaminated sites. The samples were mixed and introduced in both biotrickling filters for enrichment using MTBE as selective pressure.

RESULTS AND DISCUSSION

Culture Enrichment. After inoculation of both biotrickling filters with the mixture of contaminated soil and groundwater, various attempts were made to increase elimination of MTBE, but took about six months before a slight elimination of MTBE (about 1 g $m^{-3} h^{-1}$) could be noticed in the reactors. Since the MTBE removal was less than 5% of the incoming feed, biodegradation was confirmed by draining the scrubbing solution out of the reactors and monitoring the decrease of MTBE gas phase over time, after the reactors were closed. The MTBE initially present in the gas phase was rapidly depleted (not shown). During the experiment, carbon dioxide concentration increased, but far more than the theoretical value calculated based on the degradation of MTBE. This reflected the high energy requirement of the process culture and showed that a significant amount of secondary substrates was available in the biofilm.

Subsequently, a biofilm sample was taken from the reactors and evaluated for its ability to degrade MTBE in liquid batch cultures. High MTBE degradation rates were obtained after three one-week-long subcultures with 100 ppm MTBE.

After that, 50 to 500 ppm of MTBE could be degraded at rates of about 2 ppm per hour in liquid cultures. While the culture had a very low biomass yield coefficient, MTBE biodegradation was not subject to substrate inhibition at concentrations as high as 500 ppm. The same consortium was also capable to degrade tert-butyl alcohol (TBA), a potential metabolite of MTBE biodegradation, at a similar rate. The consortium was re-inoculated in the trickling filters hoping that the highly active suspended cells would enable better removal performance to be obtained.

Biotrickling Filter Performance after Reinoculation. Figure 1 shows the inlet and outlet concentrations of MTBE and the percentage of the MTBE degraded recovered in carbon dioxide for one of the biotrickling filters. Similar data were obtained for the other biotrickling filter and are not presented herein. Performance reached a relatively steady value at about 95% removal after 52 and 40 days, for the lava rocks and Pall ring reactors, respectively. The slight differences in startup observed between the two biotrickling filters are most probably due to the differences in surface properties of the support materials. Clearly expanded lava rocks offer many small pores easily colonized by microorganisms, whereas the polypropylene surface is not very favorable for bacterial attachment.

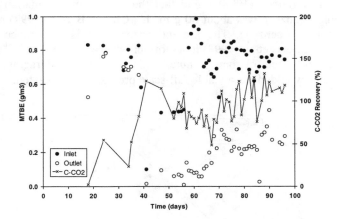

FIGURE 1. MTBE inlet and outlet gas phase concentrations and carbon dioxide recovery for reactor packed with Pall rings over the duration of the experiment. Note that time zero corresponds to the re-inoculation of the biotrickling filters.

One reason for the slow startup was the low biomass yield of the MTBE degrading process culture. The biotrickling filters only accumulated 350 g of biomass within the first 60 days of the experiment. This is a significantly lower value than biotrickling filters degrading toluene, diethyl ether or other pollutants easier to assimilate that can be subject to complete clogging by excess biomass in matter of weeks (Cox and Deshusses, 1998). The low biomass formation on MTBE was consistent with the high degree of CO_2 recovery observed in the biotrickling filter (Figure 1), which was typically 20 to 40% higher than the mineralization observed with toluene (Cox and Deshusses, 1999).

Steady-State Performance. After high removal percentage were demonstrated, MTBE loadings were increased several times to find the max performance of the systems. The inlet concentrations were kept constant and the air flow rate was increased. As reported in Figure 1 this load increase induced a significant breakthrough of MTBE after day 63. Eventually, at the higher loadings, while the removal percentage decreased, the reactors reached their maximum elimination capacity (Figure 2). In this regime, the process is kinetically limited and zero order biodegradation kinetic exists throughout the entire height of the biotrickling filter. In general, the Pall ring reactor performed a better than the lava rock one (42 vs. 50 g m^{-3} h^{-1}). In any case, the elimination capacities reached remarkably high values in the light of the fact that MTBE was considered as a recalcitrant compound until recently and considering that the maximum elimination capacity value obtained is close to those reported for the elimination of other volatile compounds in biotrickling filter. Cox and Deshusses (1999) reported a maximum elimination capacity of 70 to 85 g m^{-3} h^{-1} for toluene in a very similar system, Mpanias and Baltzis (1998) found that the maximum elimination capacity for mono-chlorobenzene was about 60 g m^{-3} h^{-1}, Oh and Bartha (1997) reported 50 g m^{-3} h^{-1} for nitrobenzene. The present performance is also much higher than reported by Eweiss et al. (1998) in a biofilter who reached 8 g m^{-3} h^{-1} for an empty bed retention time of one minute. This is very encouraging, since such performance would allow a viable full-scale process to be operated.

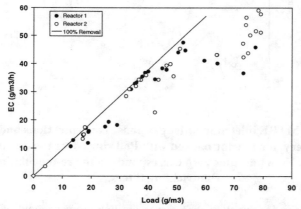

FIGURE 2. MTBE elimination capacity vs. MTBE loading for the two biotrickling filters.

Biomass Yield, Carbon Balance, and MTBE Metabolism. To date, little data exist on the mineralization of MTBE to carbon dioxide and possible formation of degradation metabolites. Several authors report that the first step in the MTBE biodegradation pathway is the cleavage of the ether bond leading to tert-butyl alcohol (TBA) and formaldehyde (Salanitro, 1995; Steffan et al., 1997). Hence both liquid and gaseous samples were carefully analyzed by gas chromatography for the presence of TBA or other metabolites. None were found. Similarly, none of the liquid samples tested positive for formaldehyde. These results are consistent with the high degree of mineralization of MTBE that was observed (Figure 1 and Table 1) and also with the results shake flasks experiments showing no formation of TBA during MTBE biodegradation (Fortin and Deshusses, 1997).

TABLE 1. Comparison of key parameters of MTBE and toluene degrading biotrickling filters.

Parameter	Biotrickling filter #1 MTBE[a]	Biotrickling filter #2 MTBE[a]	Biotrickling filter degrading toluene[b]
Period examined (days)	71 to 74 (4days)	72 to 88 (16 days)	22 to 34 (12 days)
Average load (g m^{-3} h^{-1})	76.5 ± 2.4	71.9 ± 6.5	70.7
Average elim. capacity (g m^{-3} h^{-1})	41.7 ± 3.9	50.0 ± 7.2	32.5
Average C-CO2 recovery (%)	97 ± 17	98 ± 16	51.6
Wet biomass at start (g)[c]	1050	650	4,535[d]
Wet biomass at end (g)[c]	1120	970	7,835
Biomass production rate (g$_{dw}$ m^{-3} h^{-1})[c]	4.4 ± 1.1	5.0 ± 1.2	22.3
Biomass yield (g$_{dw}$ g^{-1}$_{VOC}$)	0.105	0.0996	0.69
Specific activity (g$_{VOC}$ g^{-1}$_{dw}$ h^{-1})	(6.4 ± 0.5)×10^{-3}	(10 ± 2)×10^{-3}	(2.7 ±0.8)×10^{-3}

[a]This study; [b]compiled from Cox and Deshusses, 1999; [c]Conversion, dry biomass weight (dw) = wet biomass * 0.0538; for Cox and Deshusses (1999) conversion factor is 0.046; [d]Value for a 23.6 L biotrickling filter.

Detailed analysis of biotrickling filter operation enabled the calculation of an overall yield coefficient specific for the process culture in the biotrickling filter (Table 1). The obtained values range from 0.09 to 0.12 g$_{dw}$ g^{-1}$_{MTBE}$ which is an extremely low value compared to either other biotrickling filter studies or liquid culture data that are generally in the range of 0.4 - 0.8 g$_{dw}$ g^{-1}$_{substrate}$. Clearly, a low biomass yield is a drawback for system startup, but is advantageous over the long run, since the biotrickling filter will not be subject to clogging by overgrowing biomass. Based on the numbers of Table 1, a specific activity defined as the mass of MTBE degraded per gram of dry biomass in the biotrickling filter per hour could be calculated. This value was 5.5×10^{-3} and 11×10^{-3} g$_{VOC}$ g^{-1}$_{dw}$ h^{-1} for reactors 1 and 2, respectively. The difference between the two trickling filters is due to the higher biomass content in biotrickling filter 1 but a relatively similar performance to reactor 2. Still, the specific activities obtained are two to four times higher than those found for toluene in similar systems. This emphasizes that 1) competent MTBE degrading culture are extremely active turning over MTBE into carbon dioxide, mostly because of the difficulty to obtain sufficient energy and growth out of MTBE, and 2) in the toluene degrading biotrickling filter, a good fraction of the biomass was inactive.

CONCLUSIONS

This one of the first papers reporting an effective biotreatment process for MTBE. Still a number of questions pertaining to the biodegradation of MTBE and to the proper management of MTBE degrading cultures for remediation purposes remain unanswered. But as the number of sites contaminated with MTBE is rising, cost effective solutions for remediation are urgently needed. In this respect, the results presented herein show promising perspectives for future field implementation of biotrickling filters and for the development of effective strategies for MTBE bioremediation.

ACKNOWLEDGMENTS

Regenesis, Bioremediation Products Inc. San Juan Capistrano (CA). E.D. Reynolds (The Reynolds Group, Tustin, CA). M. Schirmer (University of Waterloo, Ont. Canada). Jack Jett (UC Riverside).

REFERENCES

Cox, H. H. J., M. A. Deshusses. 1998. *Current Opinion in Biotechnology.* 9(3): 256-262.
Cox, H. H. J., M. A. Deshusses. 1999. *Biotechnol. Bioeng.* 62(2): 216-224.
Deshusses, M. A. 1997. *Current Opinion in Biotechnology.* 8(3): 335-339.
Devinny, J. S., M. A. Deshusses, T. S. Webster. 1999. *Biofiltration for Air Pollution Control*, CRC-Lewis Publishers: Boca Raton, FL.
Eweis, J. B., E. D. Schroeder, D. P. Y. Chang, K. M. Scow. 1998. In Proceeding of the *First International Conference on Remediation of Chlorinated and Recalcitrant Compounds*. Battelle Press, Columbus, OH.
Fortin, N. Y., M. A. Deshusses. 1997. Presented at the *Annual Meeting of the American Institute of Chemical Engineers*, Los Angeles, CA, November 19, 1997.
Happel, A. M., E. H. Beckenbach, R. U. Halden. 1998. Lawrence Livermore National Laboratory. Report N° UCRL-AR-130897, Livermore, CA, 1998.
Mpanias, C. J., B. C. Baltzis. 1998. *Biotechnol. Bioeng.* 59(3): 328-343.
Oh, Y. S., R. Bartha. 1997. *J. Ind. Microbiol. & Biotechnol.* 18(5): 293-296.
Salanitro, J. P. 1995. *Current Opinion in Biotechnology.* 6(3): 337-340.
Smith, F. L., G. A. Sorial, M. T. Suidan, A .Pandit, P. Biswas, R. C. Brenner. 1998. *J. Air Waste Manage. Assoc.* 48(7): 627.
Squillace, P. J., J. S. Zogorski, W. G. Wilber, C. V. Price. 1996. *Environ. Sci. Technol.* 30(5): 1721-1730.
Steffan, R. J., K. McClay, S. Vainberg, C. W. Condee, D. Zhang. 1997. *Appl. Environ. Microbiol.* 63(11): 4216-4222.

CHARACTERIZATION OF A FILTERING MEDIUM PROPERTIES FOR THE GAS-PHASE TREATMENT OF DICHLOROBENZENES

Marie-Julie Gravel, François Roberge, Frédéric Shooner, Christophe Guy, **Réjean Samson.**
NSERC Industrial Chair in Site Bioremediation, École Polytechnique de Montréal, Montréal, Qc, Canada

ABSTRACT :

Biofiltration was evaluated for the treatment of an effluent containing gaseous dichlorobenzenes (DCBs) and naphthalene that was collected from a stripping tower used to treat contaminated groundwater at a refinery site. The filtering media consisted mainly of peat moss and demonstrated favorable physical and chemical properties, as well as a high potential of microbial growth and activity. Mineralization tests demonstrated that pH, temperature, and the presence of naphthalene in the gas phase were the factors most influencing DCB biodegradation. The pilot scale biofilter studies demonstrated the excellent potential of this technology for the treatment of gas-phase DCB-contaminated effluents. The 0.4-m^3 pilot biofilter was able to treat up to 28 g DCB/m^3media-h with a maximum efficiency of 89%. Two full-scale biofilters filled with 45 m^3 of filtering medium were also used to treat 57 m^3/h of DCB-contaminated air. Removal efficiencies obtained varied between 0 and 79%, mainly because of low DCB inlet concentrations.

INTRODUCTION

Chlorobenzenes (mono, di-, and trichlorobenzenes) are used as solvents, heat transfer agents, insect repellents, deodorants, degreasers and intermediates in dye and pesticide systhesis. Consequently, they are detected in surface water and groundwater, in sewage, and in some biological tissues, and they are identified as priority pollutants by the U.S. Environmental Protection Agency.

Chlorobenzene (CB) and 1,2-dichlorobenzene (DCB) are chemically stable and their photochemical degradations in soil and aquatic environments are limited. However, they have been mineralized under appropriate conditions in the laboratory by bacteria isolated from soil and water. Namkung and Rittman (1987) reported that biodegradation was the prevalent mechanism for the removal of CB from wastewater.

Treatment of volatile organic compounds (VOCs) in air with biofiltration has been studied for different contaminants. Corsi and Seed (1995) reported that 6 biofilters of compost, bark or peat moss, removed from 70 to 95% of benzene, toluene, ethylebenzene, and xylenes (BTEX) with rates of 6 to 64 g/m^3-h. Ergas et al. (1994) used a compost biofilter to obtain 98% removal of dichloromethane in air containing between 3 and 50 ppmv. Elimination capacity of more than 15

g/m^3-h with residence time of less than 1 minute have been reported. Oh and Bartha (1994) demonstrated that air contaminated with 0.3 g/m^3 of 1,2-DCB was treated at rates ranging from 15 to 200 g/m^3-h in a trickling filter inoculated with two bacterial consortia isolated from a contaminated industrial sludge. The water was recirculated and the pH was adjusted by an automatic pH control module using sodium hydroxyde to keep the pH in the neutral range. Because process controls such as pH adjustment and salt removal are needed on the recirculated water of the trickling air biofilter, a peat-moss type biofilter was proposed for this project.

The objective of this study was to demonstrate that biofiltration can be a suitable technology for the treatment of DCB-contaminated gaseous effluents.

MATERIAL AND METHOD

Filtering Medium. The media used in this study was obtained from a pilot scale biofilter treating a mixture of DCB. It consisted of a blend of 70% peat moss and 30% composted chicken manure, maple wood chips, and DCB-contaminated soil. The latter was used as a source of DCB-degrading microorganisms. The physical and chemical characteristics of the medium are presented in Table 1.

TABLE 1. Physical and chemical characterization of the filtering medium.

Characteristics	Measured value	Units
Total organic carbon	14	%
Total Kjeldhal nitrogen	7,200	mg/kg
Ammonia nitrogen (N base)	96	mg/kg
Soluble nitrates (N base)	5	mg/kg
Nitrites (N base)	< 1	mg/kg
Total phosphorus	2,300	mg/kg
Available phosphorous	160	mg/kg
Sulfates	44	mg/kg
pH	7,37	-
Alkalinity (in $CaCO_3$)	62,000	mg/kg
Water content	56.0	mass %
Density	220	g/L
Porosity	74.7	%

Gas-Phase Microcosms. Mineralization studies were conducted using simple batch systems in 120-ml serum bottles sealed with Mininert valves (Supelco Inc., Bellefonte, PA). A sample of 20 g of medium was added to each bottle, to leave a headspace of approximately 80 ml. The headspace amount was estimated sufficient to ensure the presence of all the oxygen needed to complete the mineralization both of the organic matter in the filtering medium and of all the added carbon. Each microcosm also included a 5-ml glass tube filled with 1 ml of 1 N potassium hydroxide (KOH) placed within the serum bottle to trap the $^{14}CO_2$. The required volume of contaminant was added by injection through the septum. After preparation, microcosms were stored at room temperature in a cabinet and

KOH samples were collected twice a day, until most of the biodegradation had occurred.

Abiotic controls also were prepared to account for the abiotic processes during the experiments. 0.04% w/w NaN_3 was added in each of the controls to kill the existing biomass in the medium.

Radioactivity of the trapped $^{14}CO_2$ was determined by liquid scintillation counting. The KOH samples were transferred to scintillation cocktail (OptiPhase HiSafe-3, Wallac, Montreal, Quebec) and analyzed using a Wallac Model 1409 scintillation counter (Wallac, Montreal, Quebec).

Pilot-Scale Biofilter The pilot scale biofilter (1.80 m high and 0.57 m diameter) was filled with 0.4 m^3 of filtering medium. The system also included a pre-humidification tower, sampling ports (for the gas-phase and the medium) and several measurement tools for the monitoring of operating conditions. Contamination of the incoming air stream was done by bubbling a secondary stream in liquid 1,2-DCB. Medium samples were collected weekly and were used to monitor the physical and chemical conditions as well as the microbiological activity in the biofilter.

The contaminated air was fed to the biofilter at a rate of 236 ± 16 L/min with a contaminant mass loading of 15.7 ± 4.0 g C/m^3-h.

Industrial-Scale Biofilter The industrial biofilter contained 45 m^3 of filtering medium. Water content and temperature in the medium were measured on-line using 4 water content reflectometers and 4 thermocouples.

Air samples from both the pilot and the industrial biofilters were analyzed on a Saturn II GC/MS equipped with a 0.32 mm x 60 m DB-1 capillary column. The carrier gas was helium (0.8 ml/min). The temperature of the column was initially maintained at -50°C for 6 minutes, then increased to 10°C at a rate of 4°C/min and finally to 160°C at 8°C/min, a temperature that was maintained for 15 minutes. Water analyses were performed on GC/MS equipped with a purge and trap system.

RESULTS AND DISCUSSION

Medium Characterization. The characterization of the filtering medium demonstrated its favorable physical and chemical properties. The low density and high porosity ensures homogeneous gas flow through the medium and avoids a high head loss. Furthermore, the presence of a variety of nutrients at relatively high concentrations allows a high potential of microbial growth and activity. Finally, the physical characteristics and chemical composition of the medium are comparable to different media characteristics found in the literature (Hodge and Devinny, 1995; Yang and Allen, 1994; Rho et al., 1994).

Influence of the Medium Characteristics on DCB Mineralization. This experiment was conducted using a factorial design studying the influence of water content, pH and temperature on 1,2-DCB mineralization. It revealed that pH and temperature both had a major influence. Acidification of the medium (from 7.5 to 6.0) caused a significant decrease in 1,2-DCB mineralization. The maximum level of mineralization reached during the experiment dropped from 60% to below 20%. Keeping in mind the risk of acidification of the filtering medium in the biological treatment of chlorinated contaminants, the influence of pH shows major consequences for the continuous operation of a biofilter. This characteristic needs close monitoring to prevent system failure from acidification of the medium.

Temperature also appeared to have an important influence on 1,2-DCB mineralization. It was observed that 25°C was the optimum temperature for mineralization. Both an increase to 40°C, and a decrease to 10°C, caused a significant decrease in the mineralization of 1,2-DCB. Water content, however, did not seem to have a significant effect on 1,2-DCB mineralization in the range of the study. Levels of mineralization for water contents of 40, 55 and 70% w/w were not statistically different.

Substrate Interaction Study. The objective of this study was to identify and characterize the interactions between 1,2-DCB, 1,4-DCB, and naphthalene. It was observed that naphthalene had a significant influence on both 1,2-DCB and 1,4-DCB mineralization. This influence was characterized by a decrease in levels of mineralization obtained and in rates of biodegradation, as well as an increase in the lag period. These observations suggest that naphtalene could act as an inhibitor for the mineralization of DCB. The opposite effect also was observed but to a much lower degree: 1,2-DCB showed a small but significant negative impact on naphthalene mineralization. Interactions between 1,2-DCB and 1,4-DCB, however, were not observed.

Pilot and Industrial Scale Operation. The pilot scale biofilter studies demonstrated excellent potential of this technology for the treatment of DCB-contaminated gas-phase effluents. It was operated over a period of 67 days (see Figure 1). During that time, two phases were observed. The first one, the start-up phase, took place from day 0 to 35 and was characterized by an adaptation of the microorganisms to the contaminant loading and a progressive drying of the medium, as well as low and variable levels of performance.

The adaptation of the microorganisms took place over the first 30 days and was characterized by an increase in the mineralization capacity for 1,2-DCB in the filtering medium. More specifically, a progressive increase in the level of mineralization and a progressive decrease in the lag period were observed.

The monitoring of the medium's physical characteristics revealed that the water content was slowly but progressively lower. Over the first 35 days of operation, its value decreased from 58.6 to 50.7% w/w. The pH and temperature variations were small; pH ranged between 7.6 and 8.7 and temperature remained between 14 and 19°C.

As can be seen from Figure 1, performances and feed conditions for this first operation phase were highly variable. Mass loading varied from 2 to 52 g DCB/m^3-h. This important variation is the result of problems encountered with the feed contamination system. The latter was replaced between day 5 and 15. Afterwards, the mass loading was more stable with a value of 32.6 ± 8.1 g DCB/m³-h. Removal efficiency and elimination capacity also were highly unstable, ranging from 0 to 50% and from 0 to 18 g DCB/m³.h, respectively.

Modifications to the operating conditions were undertaken in the second phase (day 35 to 67) to increase the performance level. These modifications were based on assumptions made to explain the poor performances obtained in the first 35 days of operation. The main assumption was that there could have been a limitation in mass transfer through the pilot biofilter. To overcome this limitation, the water content of the medium was increased by manual surface spraying over the days 44 to 51 leading to an increase in the water content from approximately 50 to 58 %w/w. This action induced a slight improvement in contaminant removal. The actual breakthrough in removal was achieved in the last days of operation, after an increase in the residence time at Day 60. The maximum elimination capacity reached was 28 g DCB/m³-h, corresponding to a removal efficiency of 89%.

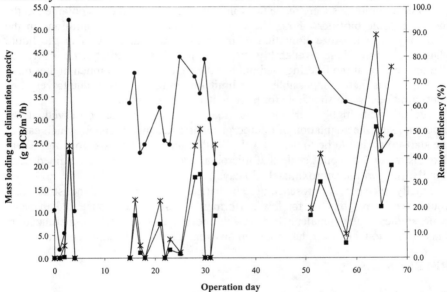

FIGURE 1. Pilot-scale removal performances

Industrial Scale Operation. Full-scale operation of two 70-m³ biofilters was studied for the treatment of air contaminated with DCBs, during 4 months. This air came from a stripping unit treating DCB-contaminated groundwater. DCB concentrations in the water stripper outlet always were under 0.5 ppb, except for one measurement at 4 ppb.

DCBs concentrations in the air inlet of the biofilters fluctuated between 100 to 1,800 ppbv for the period of the survey. DCB concentrations in the outlet of the biofilters varied between 75 and 995 ppbv and removal efficiency varied from 0 to 79%. Naphthalene also was measured in the air entering the biofilters, and concentrations between 2 and 7,800 ppbv were measured while 0 to 94% removal was obtained in the biofilters.

Water content (60% w/w) and pH (7.5) were constant during the time of the survey, but temperature decreased from 52°C to 12°C (biofilter 1) and from 46°C to 23°C (biofilter 2). These conditions are favorable for the degradation process except for temperatures lower than 20°C.

Despite the low DCBs concentrations, results showed a rise of the DCBs elimination capacity rose with an increase in the DCB inlet concentrations, even if the naphthalene concentrations were also greater at the same time. These results, combined with the mineralization tests in microcosms, demonstrated the potential of the technology for a full-scale application. Further studies will allow the optimization of the mixture content of the filtering medium for biodegradation and mechanical properties, as well as a more extensive evaluation of the interactions between dichlorobenzenes and naphthalene and an evaluation of the effect of an increased load on the overall removal results.

Three assumptions were made to explain the poor performances obtained in the industrial scale biofilter. First, the very low contaminant concentrations in the feed may have led to a limitation of the compound's dissolution in the liquid phase, lowering their availability to microorganisms. Pilot-scale operation demonstrated that the filtering medium has the capacity to treat much higher mass loadings. It is therefore possible that a limitation in mass transfer due to very low concentrations was responsible for poor contaminant removal.

Second, with highly variable mass loadings like the ones observed at the industrial site, the adaptation of microorganisms takes much longer, which causes the start-up period to be longer as well. With a more stable mass loading, it was demonstrated at the pilot scale that microbial adaptation could be obtained quite rapidly, (i.e., within approximately 30 days).

Finally, the sporadic presence of naphthalene in the feed at the industrial scale could also be responsible for low performances. It was demonstrated in batch scale studies that naphtalene had an inhibitory effect on DCB mineralization. This interaction, however, has not been investigated under dynamic conditions.

REFERENCES

Corsi, R.L. and L. Seed. 1995. "Biofiltration of BTEX: media, substrate, and loadings effects." *Environ. Progress, 14*(3):151-158.

Ergas, S.J., K. Kinney, M.E. Fuller and K.M. Scow. 1994. "Characterization of a compost biofiltration system degrading dichloromethane." *Biotechnology and Bioengineering, 44*(9): 1048-1054.

Hodge, D.S. and J.S. Devinny. 1995. "Modeling Removal of Air Contaminants by Biofiltration." *Journal of Environmental Engineering, 123*(6): 577-585.

Namkung, E. and B.E. Rittmann. 1987. "Estimating volatile organic compound emissions from publicly owned treatment works." *J. Water Pollution Control Fed, 59*: 670-678.

Oh, Y.S. and R. Bartha. 1994. "Design and performance of a trickling air biofilter for chlorobenzene and *o*-dichlorobenzene vapors." *Applied and Environmental Microbiology, 60*(8): 2717-2722.

Rho, D., P. Mercier, J.F. Jetté, R. Samson, J. Lei and B. Cyr. 1994. "Performances of a 30-L Biofilter for the Treatment of Contaminated Air with Toluene and Gasoline Vapors". *Filter Media Selection, Microbial Activity and Substrate Interactions*. GASReP Conference, Calgary, Alberta, Canada.

Yang, Y. and E.R. Allen. 1994. "Biofiltration Control of Hydrogen Sulfide. Part 1. Design and Operational Parameters." *Journal of the Air & Waste Management Association. 44*: 863-868.

MOLECULAR MONITORING OF MICROBIAL POPULATIONS DURING BIOREMEDIATION OF CONTAMINATED SOILS

DeEtta K. Mills (George Mason University, Fairfax, VA)
Kristin Fitzgerald (George Mason University, Fairfax, VA)
Patrick M. Gillevet (George Mason University, Fairfax, VA)
Carol D. Litchfield (George Mason University, Fairfax, VA)

ABSTRACT: Molecular techniques, such as DNA fingerprinting, have greatly expanded the ability to profile the complexities of natural microbial communities without dependency upon cultured-based techniques. Protocols were developed and used for the direct extraction of total soil community DNA from the original contaminated soils and from four bioreactor slurries. PCR protocols and amplification conditions were developed using prokaryotic universal primers that amplify whole community 16S ribosomal genes. PCR products were cleaved with restriction endonucleases using three different enzymes, Ava II, Hha I and Rsa I. The resulting whole community DNA fingerprints showed unique banding patterns over time as well as between control and treatments. This study has shown that molecular techniques can provide enhanced resolution and understanding of the dynamics of the microbial community and were used to monitor biotic changes during bioremediation of contaminated soils.

INTRODUCTION

Bioremediation is a viable, cost effective alternative technology for the remediation and reclamation of many petroleum-contaminated sites. Much attention is paid to the contaminant degradation. However, the microorganisms are often neglected in the studies. Many criticize the inability to assess directly the biotic processes *in situ* (Block, et al., 1993). Recently, however, molecular techniques for the elucidation and analyses of prokaryotic genes, such as 16S ribosomal genes, have greatly expanded the ability to profile the complexities of natural microbial communities. Naturally occurring base substitutions, deletions or insertions within the nucleic acid sequences can provide molecular markers that can be used to distinguish between different genomes. Cutting nucleic acids with restriction enzymes produces fragments with different numbers of bases, thus different length fragments (restriction fragment length polymorphism (RFLP)). Upon separation of the fragments by gel electrophoresis, a unique banding profile or DNA fingerprint is produced. Using these molecular technologies, our understanding of microbial community dynamics during bioremediation processes can be greatly enhanced.

Objective. One of the objectives of this bench-scale study was to develop robust molecular genetic fingerprinting techniques that could follow progressive changes in the microbial community during the bioremediation of petroleum-contaminated soils. The advantage of nucleic acid-based methods is being able to overcome the limitations and bias of traditional culture-based techniques when probing microbial communities.

MATERIALS AND METHODS

Bioreactors. Four up-flow bioreactors were inoculated with a slurry (200 grams soil/500 ml liquid) from the petroleum-contaminated soil mixture (weathered and not weathered) and operated for 31 days (Fitzgerald et al., in preparation). Briefly, three of the bioreactors continuously received a supplement of nitrogen and phosphorus (amended) while the control reactor received only water (unamended). On day seventeen, an additional spike of the weathered soil, freshly contaminated with petroleum, was added to each bioreactor and the experiment continued for additional 14 days. Samples for DNA analyses were taken eight times during the 31 days and the 1-2 grams of slurry were immediately stored at −70°C for DNA analysis.

DNA extraction, purification and amplification. The contaminated soils were used to inoculate low nutrient agar plates or trypticase soy broth and the culturable sub-community and individual isolates were collected and used as positive controls.

Total community DNA was extracted from replicate samples (500 mg/wet weight) of the original contaminated soils and from bioreactors slurry samples using the BIO 101 FastDNA™ Spin Kit for Soil (BIO 101, Inc.) with modifications. All crude DNA extracts were concentrated and further purified using Microcon 100 microconcentrators (Amicon, Inc.). The polymerase chain reaction (PCR) was performed using T*fl* DNA polymerase (Promega) and one of three sets of procaryotic universal primers that amplify 16S ribosomal genes: 8F and 1492R; 63F and 1387R (Marchesi, et al., 1998) or fluorescently-labeled 63F-NED and 1387R-6-FAM. Bovine serum albumin (fraction V) was added to the PCR reaction mix at a final volume of 0.1% to enhance PCR amplification. PCR products were purified using Wizard minicolumns (Promega) and then digested overnight at 37° C using three different restriction enzymes, Ava II, Hha I or Rsa I. The DNA fingerprints were resolved by gel electrophoresis in a 3% 3:1 Nu-Sieve (FMC Products) agarose gel stained with ethidium bromide. The image was captured and analyzed using Kodak 1D Image Analysis software. For resolution of the fluorescently-labeled terminal length fragments, samples were loaded on a 7% denaturing polyacrylamide gel with ROX 2500 internal standard, electrophoresed overnight using an ABI 373 DNA sequencing machine, and analyzed with Genescan and Genotyper analysis software (Applied Biosystems).

RESULTS AND DISCUSSION

Whole community DNA fingerprints. Using 16S rDNA primers (8F and 1492R) and conventional 3% agarose gels stained with ethidium bromide, whole community DNA fingerprints from the original contaminated soils proved to be more complex than those of the culturable soil community. When samples were cut with the endonuclease Ava II, the cultured community profile represented only 29% of the bands present in the original whole soil community profile; for Hha I, only 44% and Rsa I, only 63% (data not shown).

Slurry samples from the bioreactor experiment were also processed for whole community DNA fingerprints. Amplification products from PCR reactions using 63F and 1387R primers were digested with either Hha I or Rsa I restriction endonucleases, resolved on 3% agarose gels, stained with ethidium bromide and further analyzed. Comparisons between primer sets proved that the 63F:1387R set was much more successful in amplifying the whole community DNA than was the 8F:1492R primer set. This is consistent with published results from Marchesi, et al., 1998.

Similarity Indexes. DNA fingerprints are able to reflect similarities and differences between individuals or populations. Therefore, similarity indexes (S_{ab}) were calculated for banding profiles from analysis data compiled from gel images using the following equation (Archer and Leung, 1998):

$$S_{ab} = 2n_{ab} / (n_a + n_b)$$

where S_{ab} = the similarity index between lanes a and b

n_{ab} = the number of bands in common to both lanes

n_a and n_b = the number of bands in lane a and lane b

To better analyze the data, the bioreactor experiments were categorized into two distinct phases: Phase 1 (T_0-T_{17}), before supplementing with spiked weathered soil, and Phase 2 (T_{18}-T_{31}), after the supplement. The results of three replicate Hha I digests of whole community DNA resolved on a 3% agarose gel is shown in Table 1A & 1B. Phase 1 mapped the diversity present in the soils and the T_0 slurry made from the mixed soils. A distinct loss of bands could be seen by day fourteen especially in the unamended control. There was a unique band that was not seen previously. These population shifts in the whole community and the lack of homozygosity over time were also reflected in the calculated similarity index (Table 2A & 2B). The presence of a unique band in the T_{14} control (540-570 bp range) not previously seen in any of the original samples can be speculated to be a function of one or perhaps two main factors: (1) There was a two log difference in total numbers of bacteria (TCFUs) between the two time points ($T_0 = 10^6$, T_{14} control = 10^8, K. Fitzgerald, personal communication) which theoretically would yield more total genomic DNA as well as change the ratio of DNA templates in the community. The PCR process can now more efficiently amplify this previously rare template. (2) By T_{14}, the nutritional environment of the bioreactors had changed (K. Fitzgerald, personal communication). The population represented by the 540-570 band could be the only population that successfully adapted to low nutrient conditions or could use the more recalcitrant compounds. We cannot distinguish between these two alternatives.

Band 540-570 was still present in the T_{17} control sample that was taken two hours after the supplemental spiked soil was added. This population seemed to remain stable while other populations that had previously disappeared from the profile were adapting to the newly added nutrients (i.e., the Arabian light crude oil). At T_{21}, the 540-570 band was absent, a possible competitive exclusion or a lack of availability of a preferred carbon source (a possible by-product). The band reappeared by day thirty-one when the nutritional environment was again limited. Replicate DNA extractions, PCR runs, restriction digests and fingerprint gels consistently reproduced the same community DNA fingerprints.

Table 1: A: Phase 1 (T0-T17). B: Phase 2 (T18-T31): The presence or absence of DNA restriction length fragments showed a distinct fingerprint profile over time and between the original soils, the control and treatments.

A: Phase 1

Base Pair Range	Un-weathered soil	Mixed soils	T0 Slurry	T14 Control	T14 Treated	T17 Control	T17 Treated
1050-1130	--	++	++	--	--	--	++
920-940	++	++	--	--	--	--	--
600-640	++	++	--	--	--	--	--
540-570	--	--	--	++	--	++	--
450-475	++	++	++	--	--	--	--
350-405	--	--	++	--	++	++	++
295-330	++	++	--	++	++	++	++
270-286	++	++	++	--	++	--	++
> 200	++	--	++	--	++	--	++

B: Phase 2

Base Pair Range	Un-weathered soil	Mixed soils	T0 Slurry	T21 Control	T21 Treated	T31 Control	T31 Treated
1050-1130	--	++	++	--	--	--	--
920-940	++	++	--	--	--	--	--
600-640	++	++	--	--	--	--	--
540-570	--	--	--	--	--	++	++
450-475	++	++	++	--	--	--	--
350-405	--	--	++	++	++	--	++
295-330	++	++	--	++	++	++	++
270-286	++	++	++	++	++	++	++
> 200	++	--	++	++	++	--	++

++ represents the presence of a band; -- represents the absence of a band

Since the TCFUs remained fairly constant during Phase 2, the changing fingerprint patterns were most likely due to changing genomic ratios of the populations responding to the bioremediation processes than in simply total cells present in the samples.

Comparison of Detection Techniques. Computer analysis of most of the published bacterial rDNA sequences have shown that Hha I endonuclease can produce a total of 252 different restriction fragments between 50 and 450 nucleotides in length. Rsa I can produce 231 varying length fragments (Brunk, et al., 1996). Therefore, a protocol designed to follow whole community microbial dynamics should be sensitive enough to allow detection of the widest range of restriction site polymorphisms. Resolving faint bands, especially below 200 nucleotides in length, on ethidium bromide stained agarose gels proved

Table 2: Similarity indexes for Phase 1 (A) and Phase 2 (B) generally showed a lack of similarity between control and treatment banding profiles as well as over time.

A: Phase 1

	Un-weathered soil	Mixed soils	T0 Slurry	T14 Control	T14 Treated	T17 Control	T17 Treated
Un-weathered	1.00	0.83	0.55	0.25	0.60	0.22	0.55
Mixed soils	--	1.00	0.55	0.25	0.40	0.22	0.55
T0 Slurry	--	--	1.00	0.00	0.67	0.25	0.80
T14 Control	--	--	--	1.00	0.33	0.80	0.29
T14 Treated	--	--	--	--	1.00	0.57	0.89
T17 Control	--	--	--	--	--	1.00	0.50
T17 Treated	--	--	--	--	--	--	1.00

B: Phase 2

	Un-weathered soil	Mixed soils	T0 Slurry	T21 Control	T21 Treated	T31 Control	T31 Treated
Un-weathered	1.00	0.83	0.55	0.60	0.60	0.44	0.55
Mixed soils	--	1.00	0.55	0.40	0.40	0.44	0.36
To Slurry	--	--	1.00	.067	0.67	0.25	0.60
T21 Control	--	--	--	1.00	1.00	0.57	0.89
T21 Treated	--	--	--	--	1.00	0.57	0.89
T31 Control	--	--	--	--	--	1.00	0.75
T31 Treated	--	--	--	--	--	--	1.00

difficult to do with great accuracy even with the digital imaging hardware and software that were available. One of the great advantages of the fluorescent method is its ability to accurately resolve band sizes especially below 500 base pairs in length. In addition, each lane includes an internal standard, thereby eliminating any variation due to pipetting errors or matrix differences within a gel. Restriction digests were resolved on both agarose gels and polyacrylamide gels. Preliminary data from the fluorescent PCR terminal fragment length fragments in this study showed much higher resolution of all fragments below 500 base pairs in length eliminating the ambiguity of determining fragment size that was inherent in the agarose gels. The same changing trends in fingerprint profiles were once again demonstrated (data not shown).

CONCLUSION

This study was a part of a larger study that used more traditional-based culture and biochemical techniques to follow the population dynamics of the bioremediation of contaminated soils (K. Fitzgerald, et. al., in preparation). In general, the results from these other methods showed surprising stability in the community over time and between controls and treatments. Resolution at the genomic level, however, showed definitive changes and distinct DNA fingerprint profiles throughout the bioreactor experiment. Differences were also reflected between the unamended control samples and the amended treatment samples. As the nutritional status within the bioreactors changed over time, so did the DNA profiles. The nutrient amended samples always had a more complex DNA profile than did the controls. There was also less petroleum remaining in the treatment reactors than in the controls (K. Fitzgerald, personal communication). These differences in DNA profiles may reflect the beneficial effects of nutrient amendments for enhancing the degradation of contaminants. Techniques using fluorescent primers and state-of-the-art DNA analysis equipment has further enhanced the resolution of the DNA fingerprints and can give a more accurate picture of the microbial community (D. Mills, et. al., in preparation). Either fingerprinting approach will most likely underestimate the true diversity of natural microbial communities but DNA fingerprinting has proven a useful tool to monitor and further our understanding of microbial population dynamics during bioremediation.

REFERENCES

Archer, E. S., and F. C. Leung. 1998. "Computer Program for Automatically Calculating Similarity Indexes from DNA Fingerprints." *BioTech. 25*(2): 252-254.

Block, R., H. Stroo, and G. H. Swett. 1993. "Bioremediation--Why Doesn't It Work Sometimes?" *Chem. Engr. Prog. August*: 44-50.

Brunk, C. F., Erik Avaniss-Aghajani, and Clifford A. Brunk. 1996. "A Computer Analysis of Primer and Probe Hybridization Potential with Bacterial Small-Subunit rRNA Sequences". *Appl. Environ. Micro.62*(3): 872-879.

Fitzgerald, K. D. K. Mills, and C. D. Litchfield,. 1999 "Culture and Biochemical Changes in Bioremediated Aged Petroleum Contaminated Soils." (in preparation)

Marchesi, J. R., T. Sato, A. J. Weightman, T. A. Martin, J. C. Fry, S. J. Hiom, and W. G. Wade. 1998. "Design and Evaluation of Useful Bacterium-Specific PCR Primers That Amplify Genes Coding for Bacterial 16S rRNA." *Appl. Environ. Micro.64*(2): 795-799.

Mills, D. K., K. Fitzgerald, P. Gillevet, and C. Litchfield, 1999. "Molecular Monitoring of Microbial Communities during Bioremediation" (in preparation).

FERMENTATION OF ORGANIC SOLID WASTES AS A SOURCE OF RENEWABLE ENERGY

Ferhan Sami ATALAY (Ege University, Bornova, Izmir, Turkey)

Ayşe Hilal Yılmaz (Ege University, Bornova, Izmir, Turkey)

ABSTRACT: As a result of anaerobic fermentation of biodegradable organic wastes, a gas mixture containing methane and carbon dioxide mainly, is produced. The organic complex solid wastes that might be harmful for human can be treated using the anaerobic fermentation process producing useful energy and also fertiliser. In this study, methane production from six different solid wastes was investigated using six small batch laboratory fermenters each of that had a volume of 10 liters and a pilot scale fermenter which had a volume of 0.96 m^3. All experiments were performed at 35°C, and the results were applied to the different microbial kinetics that are indicated in literature.

INTRODUCTION

Biogas is the name given to gas mixture produced by fermenting waste, particularly sewage in anaerobic conditions. This process is an alternative method of waste disposal to landfill or conventional sewage plants. Waste is incubated with suitable bacteria in a digester in the complete absence of air (an anaerobic fermenter). In sewage treatment plants using anaerobic fermentation, methane is often used as a power source for the plant itself. The process is also called "anaerobic digestion" (Bains, 1993).

The bacteria responsible for generating methane from waste are the methanogenic bacteria, an unusual group which can turn a limited number of carbon substrates into carbon dioxide and methane, to break the waste down into things that the methanogens can eat requires other bacteria. Thus an anaerobic digester needs specialized population of bacteria to work well (Bains, 1993).

Biogas is simply the fermentation of wastes by different microorganisms. So, the parameters that affect growing and producing of microorganisms, also affect biogas production rate. These effects are: microbial community, temperature, substrate biodegradability, pH, and mixing (Atalay, 1984).

Objective. The objective of this study is the determination of gas production rates and methane percentages by anaerobic fermentation of six different organic solid wastes.

MATERIALS AND METHODS

The laboratory system consisting primarily of a heating room, heating and control elements, fermenters, agitation system, and gas counter and the pilot scale system (Figure 1 and Figure 2), and also the procedure applied were presented in detail elsewhere (Yılmaz, 1998).

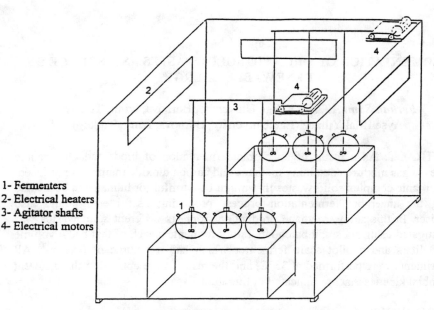

1- Fermenters
2- Electrical heaters
3- Agitator shafts
4- Electrical motors

FIGURE 1. Laboratory System

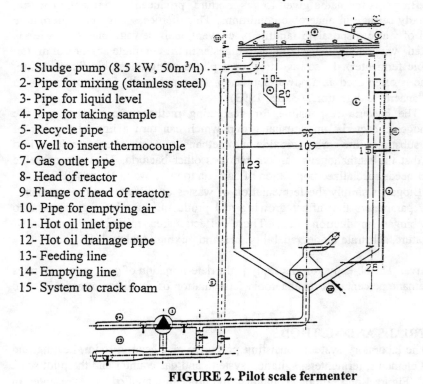

1- Sludge pump (8.5 kW, 50m^3/h)
2- Pipe for mixing (stainless steel)
3- Pipe for liquid level
4- Pipe for taking sample
5- Recycle pipe
6- Well to insert thermocouple
7- Gas outlet pipe
8- Head of reactor
9- Flange of head of reactor
10- Pipe for emptying air
11- Hot oil inlet pipe
12- Hot oil drainage pipe
13- Feeding line
14- Emptying line
15- System to crack foam

FIGURE 2. Pilot scale fermenter

During the experiments, for six different organic solid wastes, daily gas production and their concentrations were recorded. In order to obtain the initial substrate and micro-organisms amounts the method of the determination of dry solid and its organic content was used and the procedure was presented in detail elsewhere (Yılmaz, 1998). The specifications of organic solid wastes are given in Table 1.

TABLE 1. Specification of organic solid wastes

Solid Name	Dry Solid (%)	Organic Dry Solid (%)
Water Hyacinths	7.53	81.78
Food Waste	18.0	65.9
Cotton Stalk	80.18	98.9
Wastewater	1.32	85.14
Harvested grass	40.03	73.48
Corn Stalk	97.97	91.72

RESULTS AND DISCUSSION

The gas production and methane percentage changes with respect to time for each solid waste are given in the following figures (Figure 3, 4, 5, 6, 7, 8).

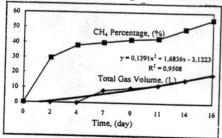

FIGURE 3. For water hyacinths **FIGURE 4. For food waste**

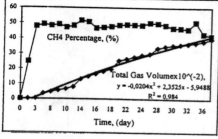

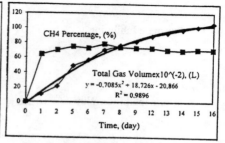

FIGURE 5. For corn stalk **FIGURE 6. For wastewater**

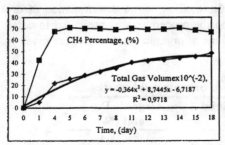

FIGURE 7. For cotton stalk   **FIGURE 8. For harvested grass**

Table 2. The derived model equations

NAME	MODEL EQUATIONS
Monod Equation	$$\frac{dG}{dt} = \frac{\mu_m(S_0 - \alpha G)(X_0 + Y\alpha G)}{Y\alpha(K_s + S_0 - \alpha G)}$$
Monod Equation With Decay Rate	$$\frac{dG}{dt} = \frac{1}{Y\alpha}\left[\frac{\mu_m(S_0 - \alpha G)}{K_s + S_0 - \alpha G} - b\right](X_0 + Y\alpha G)$$
Contois Equation	$$\frac{dG}{dt} = \frac{1}{Y\alpha}\left[\frac{\mu_m(S_0 - \alpha G)(X_0 + Y\alpha G)}{BX_0 + S_0 + \alpha G(K - 1)}\right]$$
Contois Equation With Decay Rate	$$\frac{dG}{dt} = \frac{1}{Y\alpha}\left[\frac{\mu_m(S_0 - \alpha G)}{BX_0 + S_0 + \alpha G(K - 1)} - b\right](X_0 + Y\alpha G)$$
Substrate Inhibition	$$\frac{dG}{dt} = \frac{\mu_m}{Y\alpha}\left[\frac{X_0 + Y\alpha G}{1 + \dfrac{K_s}{S_0 - \alpha G} + \dfrac{S_0 - \alpha G}{K_i}}\right]$$
Substrate Inhibition With Decay Rate	$$\frac{dG}{dt} = \frac{1}{Y\alpha}\left[\frac{\mu_m}{1 + \dfrac{K_s}{S_0 - \alpha G} + \dfrac{S_0 - \alpha G}{K_i}} - b\right](X_0 + Y\alpha G)$$

In order to design the biogas fermenters, microorganism growth kinetics, substrate consuming and gas production rate models should be determined. To develop the correlations between substrate concentration and gas production rate specific growth rate of microorganisms should be combined with the substrate material balance equation. Change of gas production with respect to time for six specific growth rate equations were derived and they are given in Table 2 (Yılmaz, 1998).

Where G= total gas production (g) ; K_i = inhibition coefficient
      K_s= saturation amount (g)   ;   S= substrate amount (g)
      X= microorganism amount (g) ;   Y= growth yield constant
      b= decay rate constant (day^{-1}) ;   t= time (day)
      μ= specific growth rate of micro-organisms (day^{-1})
      μ_m= maximum specific growth rate of micro-organisms (day^{-1})

Results of the experiments were applied to six different model representing some of kinetic parameters by using non-linear regression technique and the best fitting model relating the specific growth rate of microorganisms for each solid waste was determined. The results of the modelling studies and the kinetic parameters obtained from the best model equation for each solid waste have been presented in Table 3.

TABLE 3. Best fitting models and obtained parameters for each solid waste

SOLID WASTE	MODEL	PARAMETERS							
		μ_m	Y	α	K_s	b	K	B	K_i
Water Hyacinths (from lagoon)	Substrate Inhibition	0.1	0.16	5.6	1	-	-	-	1.9
Food Waste (from restaurant)	Substrate Inhibition With Decay Rate	0.7	0.1	7	1	0.08	-	-	54.8
Cotton Stalk (from field)	Substrate Inhibition With Decay Rate	0.7	0.04	7	1	0.02	-	-	1.59
Wastewater (from olive plant)	Contois Equation With Decay Rate	0.8	0.82	0.8	-	0.23	0.25	1	-
Harvested Grass	Substrate Inhibition With Decay Rate	0.7	0.24	2.86	1	0.01	-	-	2.02
Corn Stalk (from field)	Substrate Inhibition With Decay Rate	0.7	0.08	1.41	1	0.25	-	-	4.36

Also total gas production rates obtained from experiments and derived model equations were compared, and results have been illustrated for six organic solid waste in detail elsewhere (Yılmaz, 1998). For corn stalk this graph is presented in Figure 9.

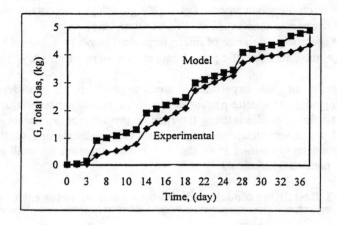

FIGURE 9. Comparison of experimental and modeling results for corn stalk

ACKNOWLEDGEMENT
This project AVI-94-0005, entitled on "Integrated Concept For The Fermentation Of Sewage Sludge And Organic Solid Waste As A Source Of Renewable Energy And For The Use Of The Fermented Product As A Hygienic Fertiliser" is supported by European Community.

REFERENCES

Atalay, F.S. 1984. "Determination of Biogas Production Conditions and Plant Design." PhD. Thesis, Ege University, Izmir, Turkey.

Bains, W. 1993. *Biotechnology From A to Z.* Oxford University Press, N.Y

Yılmaz, A.H. 1998. "Fermentation of Organic Solid Wastes as a Source of Renewable Energy andthe Use of the Product as Fertiliser." M.S. Thesis, Ege University, Izmir, Turkey

CONDITIONS AND EFFECTS OF MICROBIAL SOIL REMEDIATION IN SOLID-STATE BIOREACTORS

Jürgen Scholz and Jörg Schwedes (Institute of Mechanical Process Engineering, Technical University of Braunschweig, Germany)
Bernd G. Müller and Wolf-Dieter Deckwer (Department Bioengineering, National Research Center for Biotechnology GBF, Braunschweig, Germany)

ABSTRACT: Transport limitations cause problems like a lack of oxygen which are possible reasons for slow microbial degradation of hydrocarbons in native soils. Mixing the soil can enable sufficient oxygen and nutrient supply of microorganisms causing a higher microbial activity. To avoid an agglomeration of the moist soil a mechanical stress for deagglomeration has to be applied, leading to further increasing of microbial soil activity if the moisture content is above the plastic limit. Even a short soil treatment shows effects lasting longer than the treatment itself. The degradation of hydrocarbons can be positively influenced by both single and periodic application of mechanical stress. Possible types of solid-state bioreactors are rotary drums and siloreactors. To justify the use of solid-state bioreactors the soil has to be treated at a moisture content with maximum microbial activity. But at this moisture content the physical properties like agglomerate strength, gas permeability and flowability are worst. In spite of that the reactors can be operated without trouble by knowing and considering these properties.

INTRODUCTION

Hydrocarbon contaminants in soils can often be removed by microbial soil remediation techniques. The lack of oxygen in deeper soil layers as well as the poor availability of contaminants are the main reasons why the aerobic degradation processes are slow or do not work at all. The target of treating contaminated soils in solid-state bioreactors is an enhanced degradation, achieved by:

- mixing the soil in order to distribute and homogenize nutrients, contaminants and microorganisms
- opening fine pores formerly inaccessible for microorganisms
- mixing the soil to supply more oxygen
- adjusting optimal conditions for microbial metabolism (temperature, moisture content, nutrients, etc.)
- preventing escape of volatile toxic compounds
- using the bioreactor as a tool for soil remediation with genetically manipulated microorganisms or other special applications

The microbial activity of the soil depends on several influencing factors, e. g. moisture content, soil bulk density and height of soil layers. In order to achieve a high microbial activity these influences must be known and considered. Figure 1 shows the dependence of microbial activity (measured as soil respiration) on relative moisture content of the soil. The parameter for the moisture content is the field capacity fc. This is the amount of water held by soils after excess water has drained by gravity after a complete saturation of the soil with water.

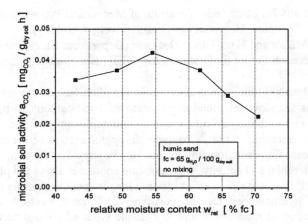

FIGURE 1. Influence of soil moisture content on microbial soil activity

Figure 1 shows that the moisture content optimal for microbial activity is in the range of about 50 % fc. This is independent of the soil type. Lower amounts of water lead to a limitation of the uptake of soluble nutrients and contaminants by the microorganisms. High amounts of water prevent a sufficient oxygen supply by saturation of the soil pore volume with water, which has a much higher diffusion resistance for the oxygen than air has. Both, increasing soil height as well as increasing soil bulk density, lead to a decreasing microbial soil activity. This is due to the worse oxygen supply of microorganisms by the increasing diffusion resistance because of the lower porosity and the longer diffusion distance, respectively.

To justify the high costs for the apparatus and the energy input when operating a solid-state bioreactor (e. g. in comparison to heap sanitation techniques as one of the usual sanitation techniques) the soil has to be treated under optimal conditions. This means an optimal moisture content, low soil heights, if there is no mixing, and prevention of consolidation of soil in the case of mixing.

PHYSICAL SOIL PROPERTIES

Similar to microbial soil activity the physical soil properties are mainly influenced by the moisture content. At a moisture content of about 50 % fc a number of physical soil properties change significantly as shown in Figures 2 and 3. The investigations described below concern the gas permeability and the plasticity of moist soils. The gas permeability was measured as pressure drop Δp in a column filled with soil of height h_s, which was permeated by air with different velocities $v_{air,0}$. The permeability resistance k is calculated from

$$k = \frac{\Delta p}{h_s \cdot v_{air,0} \cdot \eta_{air}}$$

The plastic limit w_p was determined according to the method of Atterberg (DIN, 1996), where a moist cylindrical soil sample (diameter 3 to 5 mm) is kneaded on a water absorbing material until it breaks. The plastic limit marks the moisture content at the turn-over from crumbly to plastic properties.

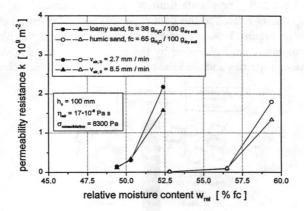

FIGURE 2. Influence of moisture content on gas permeability

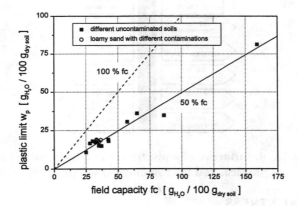

FIGURE 3. Influence of moisture content on soil plasticity

At low relative moisture contents w_{rel} (i. e. $w_{rel} < 50\ \%$ fc) the water forms isolated bridges between the soil particles. Air passes easily through the bulk both by diffusion and by convection (Figure 2). The permeability resistance is low. When soil is kneaded at this moisture content the bridges are disrupted. Thus the soil has a crumbly structure. At relative moisture contents higher than 50 % fc most pores are filled with water and the water films become coherent. This moisture content corresponds to a saturation ratio of water related to pore volume of 0.8 to 0.9. Air-filled pores are isolated and do not form a connected system. Therefore, the permeability resistance increases rapidly (Figure 2). When the moist soil is kneaded the water films still remain coherent. The particles may slide off each other causing the plasticity of the soil. The plastic limit w_p of uncontaminated and contaminated soils (moisture content related to the field capacity) does not depend on the type of soil, as shown in Figure 3. Apart from the permeability and the plasticity other physical soil properties are influenced by the moisture content. Figure 4 shows an overview of these properties and the kind of influence (Scholz and Schwedes, 1998).

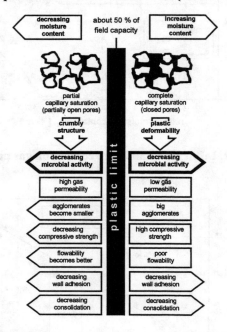

**FIGURE 4. Influence of moisture content (saturation ratio)
on physical and microbiological soil properties**

MECHANICAL STRESS

In order to ensure a sufficient oxygen supply mixing of the soil is suitable, but the simultaneously occurring agglomeration of the moist soil has to be avoided. Therefore, mechanical stress has to be applied. Possible reactors are rotary drum reactors with grinding beads and siloreactors with external conveying and recirculation (Scholz et al., 1998). The mechanical stress in the rotary drum reactor is caused by the grinding beads. A mixing without comminution of soil particles is achieved by adjusting an optimal moisture content (slightly above the plastic limit). The siloreactor including the discharge and the conveying equipment has to be designed carefully, because the flow properties of soils at moisture contents with maximum microbial activity are very poor as shown in Figure 4. Fundamentals of silo design with regard to prevention of flow problems like arching and core flow are described elsewhere in detail (Schwedes, 1996). The mechanical stress is applied to the soil mainly by the discharge equipment. An aeration of the soil takes place when the soil is on the surface of the silo content.

Several different effects with different duration can be observed after the application of this mechanical stress, as shown in Table 1.

TABLE 1. Effects of a single short application of mechanical stress

effects	influence on microbial activity	duration
aeration	increasing	few hours
release of organic compounds	increasing	< 1 day
local mixing	increasing	< 1 week
influence on number and metabolism of microorganisms	change	< 1 week
consolidation	decreasing	continuous
degradation of contaminants	**change**	**< 1 week**

The effects of the mechanical stress show a lot longer duration compared to the duration of the mechanical stress itself, but the duration of the effects is limited. The better oxygen supply is a positive effect of the treatment in the solid-state bioreactor. Whether the application of mechanical stress will show further positive effects or not mainly depends on the moisture content, as shown in Figure 5, where the microbial activity is observed for 10 days after the application of mechanical stress. The soil respiration and the degradation of contaminants after the application of mechanical stress decrease when the moisture content w is adjusted below the plastic limit w_p. Only moisture contents above the plastic limit show positive effects of a single application of mechanical stress.

Due to the limitated duration of the effects a periodical mixing / application of mechanical stress is necessary. This results in a rising consolidation of the soil. The shorter the treatment intervals are, the faster the consolidation is. If the moisture content is above the plastic limit, short treatment intervals are necessary because of the poor oxygen supply, and the rising consolidation will compensate the positive effects of a single application of mechanical stress soon. Thus, the treatment intervals have to be chosen carefully.

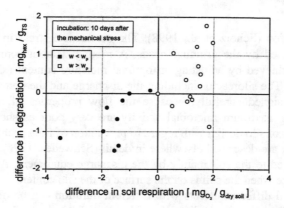

FIGURE 5. Influence of moisture content during the application of mechanical stress on soil respiration and degradation of contaminants

SUMMARY

The treatment of contaminated soils in solid-state bioreactors in order to enhance the microbial degradation is related to problems of handling the moist soil. This is due to the contrary behaviour of microbial and physical soil properties, i. e. the physical soil properties are worst just at the moisture content with maximum microbial activity, corresponding to the plastic limit. The effects, which cause the changes in degradation are different, like better aeration, homogenization, release of organic compounds as positive effects and consolidation as a negative effect. All effects can be achieved by a short soil treatment. Most effects show a limited duration which is a lot longer than that of the treatment itself. Thus, a periodic treatment with intermission times up to a few days is suitable in order to achieve a high average microbial activity. In summary the positive effects decrease with increasing duration of sanitation due to the consolidation. Nevertheless, solid-state bioreactors with mechanical stress are a suitable tool for microbial soil remediation.

REFERENCES

DIN 18122, part 1, Beuth Verlag, Berlin 1996

Scholz, J., J. Schwedes. 1998. "Use of Solid-State Bioreactors for Microbial Degradation of Organic Contaminants in Soils." *6. Int. Conf. on Bulk Materials Storage, Handling and Transportation, Wollongong, Australia, September 28–30*

Scholz, J., B. G. Müller, J. Schwedes, W.-D. Deckwer. 1998. "Use of a Rotary Drum Reactor with Grinding Beads for Microbial Soil Remediation." *Chem. Eng. Technol. 21 (6): 479-483*

Schwedes, J. 1996. "Measurement of Flow Properties of Bulk Solids." *Powder Technology 88 (3): 285-290*

AEROBIC WASTE COMPOUND DEGRADATION IN PACKED BED REACTORS

Jan Paca, Pavel Weigner, Bohumil Koutsky (all: Inst. Chem. Technol., Prague, Czech Republic), Aline V. Metris, A. Mark Gerrard (both: Univ. Teesside, Middlesbrough, UK), Katrin Frankenfeld and *Hans-Peter Schmauder* (both: fzmb, Bad Langensalza, Germany)

ABSTRACT: Immobilized biocatalysts were applied in different packed bed reactors to investigate their use for application in degradation of volatile waste compounds. The materials for the immobilization were inorganic (e.g. silicates, slate, glasses) and organic (e.g. polyurethanes, other foams). The influences of surfactants, co-substrates etc. were also examined. The microorganisms applied were of a natural origin, selected and stabilized for stable and constant degradation. Mixed cultures as well as strains with a known potential for the degradation of the waste compounds (such as styrene, TCE, BTEX) were used. These strains or mixed cultures were characterized (e.g. physiological activities, stability), as necessary. The reaction systems were characterized from the technological (e.g. flux rates, steady states, stability, mean retention times, loading rates, number of model pollutants) as well as modelling point of view.

INTRODUCTION

Volatile organic compounds (VOCs) in the air are of increasing concern. The monoaromatic hydrocarbons known as BTEX (benzene, toluene, ethylbenzene, and xylenes) are an important example. Their presence in air, soil and ground water is a widespread problem because of the leakage of underground petroleum storage tanks and spills at petroleum production wells, refineries, pipelines, and distribution terminals. Many states have established clean up standards for these chemicals because of their carcinogenic potential (Dean, 1985; Fishbein, 1985). Biodegradation of BTEX under aerobic conditions is well known. However, the main problem occuring with the decontamination is the availability of oxygen in soil and sediments. Another problem is the preparation and maintenance of a cell population with high degradation activity for BTEX. If a mixture of BTEX is to be degraded under aerobic conditions a microbial consortium has usually been used. For the complete degradation of a mixture of BTEX, a hybrid strain, *Pseudomonas putida*, was constructed recently (Lee et al., 1994). The annual quantity of BTEX produced is about 11.1 million t used as industrial solvents and/or feedstocks for synthesis (Reisch, 1992).

Styrene is an important bulk product produced in large quantities. Its major producers are the USA, Germany, and Japan (Tossavainen, 1978). In 1980, in the USA alone, its production was $3.64*10^6$ t (Fu and Alexander, 1992). In 1981, the styrene emissions in The Netherlands were estimated to be

1,440 t (0.5% of the total amount of styrene produced and processed) (Rijksinstitut, 1986). Styrene is suspected to be carcinogenic (cf. mutagenic styrene epoxide which is formed in the liver) (Vainio et al., 1992).

MATERIALS AND METHODS

Bioreactor. The experiments were performed in packed bed bioreactors. Either peat and polystyrene balls (in the ratio 50% to 50% of weight) or peat, bark and wooden chips (68% to 12% to 20% weight) were used as packings. The bed height was 250 mm, the temperature of operation was 20-23^0C.

The packed bed reactors were described earlier by Paca et al. (1998). The inlet air was continually contaminated by pure components or mixtures of xylene, toluene or styrene. The loading of pollutants were carried out by changing either their inlet gas concentrations or the rate of the volumetric gas flow. The pH of the media containing mineral salts and essential nutrients was approximately 6.0 - 7.0.

Carriers. In some experiments, carriers of an inorganic as well as organic origin were used, e.g. the porous materials *Terraperl S* or *Öl-Ex hart* (Schmauder et al. 1997), expanded slate or perlite as well as polystyrene. These materials have a large surface area for the immobilization of the microorganisms.

Microorganisms. Mixed microbial populations were isolated from long-term contaminated areas and were taxonomically determined by Culture Collections of Microorganisms as *Pseudomonas*, *Comamonas*, *Bacillus* sp. (Paca et al. 1998; Schmauder et al. 1997).

Analyses. Analyses were performed by hplc, oxygen electrodes and Infralyt 4 (waste compounds, metabolites, O_2 and CO_2 in the air) (Paca et al. 1998; Schmauder et al. 1998)

RESULTS AND DISCUSSION

Styrene Degradation in Biofilters. The start-up periods with different packings were compared. The packings mentioned above, as well as inorganic materials like *Terraperi S* or *Öl-Ex hart* and expanded slate, were used.

The results proved that a higher quantity of peat and the use of materials with porous and hydrophilic surfaces resulted in a higher removal efficiency of styrene.

Degradation of Xylene and Toluene Mixtures. The effects of different pH values and structural properties of packings were tested during the start-up period. As an example, packing materials with peat, bark and wooden chips in the ratio 50% to 40% to 10% were used. Other parameters were the same as mentioned above.

The results revealed that a faster increase of xylene removal is achieved when the bed has a higher pH value. The specific surface area, which is available for cell immobilization, significantly influences the removal efficiency of both pollutants. For details, see Figure 1 - Figure 4.

Modelling of the Reactions. The degradation of xylene and toluene in a bench-scale biofilter was studied. The following effects on the biofilter performance were considered:
- Start-up procedure followed by a long-term loading with different packings (polystyrene or perlite and compost)
- Effect of volumetric gas flow rate (mean retention time)
- Microbiological analysis of immobilized population
- Effect of bed height during start-up operation with a following loading

The packing used was compost with Perlite or polystyrene in a ratio 80% to 20% by volume. Biodegradation conditions were: temperature of 20-22^{0}C and pH value of the bed of 6.0. The waste gas, consisting of humidified air with a mixture of xylene and toluene in the ratio 50% to 50% (volume), was used. The biofilter loading was carried out by changing either the inlet gas concentration of pollutants or the superficial gas velocity. The biofilter used had a bed height of either 280 mm or 810 mm.

Biofilters are complex devices to model. When they operate at steady state conditions, the simplest models to apply assume a rate of biological degradation per unit *time* and unit *volume* of bed. The dependence on concentration is often assumed to follow a Monod form which may be extended to allow for substrate inhibition etc.

A more fundamental approach is to model the kinetics per unit *mass* of biologically active cells and to allow for diffusional effects in the liquid layer around the damp, solid carrier. These more rigorous models soon lead to sets of non-linear ordinary and partial differential equations having to be solved.

When the biofilter is run in an unsteady state condition, the model has to be in the form of partial differential equations because concentrations depend on time, the position in the bed and the depth into the particle. The nature of the unsteady state can vary. For example, we could envisage periods of high or low flow rates (or inlet concentrations). Alternatively, there could be periods of starvation when the microorganisms have none of their normal substrate to consume. For the latter case, we found a simple empirical model (Gerrard et al. 1997) could adequately represent the changes in the measured degradation efficiency.

REFERENCES
Dean, B. J. 1985. ,,Recents Findings on the Genetic Toxicology of Benzene, Toluene, Xylenes and Phenols." *Mutation Research* 154, 153-181.

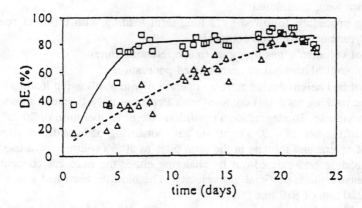

FIGURE 1. Changes of the degradation efficiency for xylene during the adaptation period. △ , dashed line - pH 3.5. □, solid line - pH 6.5.

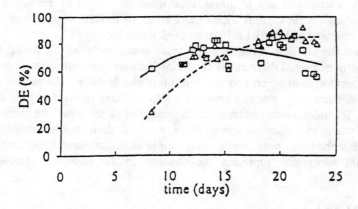

FIGURE 2. Changes of the degradation efficiency for toluene during the adaptation period. △ , dashed line - pH 3.5. □ , solid line - pH 6.5.

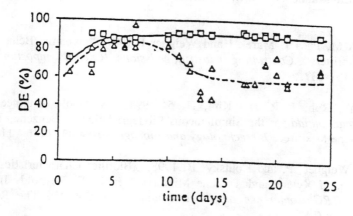

FIGURE 3. Changes of the degradation efficiency for xylene during the adaptation period. Mean bark particle size: △ - 15 - 25 mm, □ - 0.5 - 5 mm

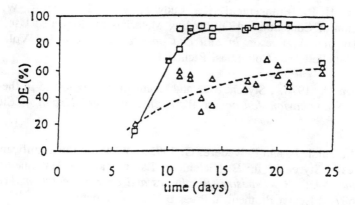

FIGURE 4. Changes of the degradation efficiency for toluene during the adaptation period. Mean bark particle size: △- 15 - 25 mm, □ - 0.5 - 5 mm

Fishbein, L. 1985. ,,An Overview of Environmental and Toxicological Aspects of Aromatic Hydrocarbons. II. Toluene." *Science of the Total Environment* 42, 267 - 288.

Fu, M. H. and Alexander, M. 1992. ,,Biodegradation of Styrene in Samples of Natural Environments." *Environmental Science and Technology* 26, 1540 - 1547.

Gerrard A M, Paca J, Marek J and Weigner P 1997. "Dynamic Behaviour of Starved Biofilters", *Controlling industrial emissions, IChemE symposium series, no 143,* p 131-137.

Lee, J. Y., Roh, J. R. and Kim, H. 5. 1994. ,,Metabolic Engineering of *Pseudomonas putida* for the Simultaneous Biodegradation of Benzene, Toluene and p-Xylene Mixture." *Biotechnology and Bioengineering* 43, 1146 - 1152.

Paca, J., Weigner, P. and Koutsky, B. 1998. ,,Biofilter Characteristics during Degradation of Xylene and Toluene Mixtures." In F. E. Reynolds Jr. (Ed.) 1998 *USC-TRG Conference on Bioremediation,* pp. 249-256 The Reynolds Group, Tustin CA.

Reisch, M. 5. 1992. ,,Top 50 Chemicals Production Stagnated Last Year." *Chemical and Engineering News* 70 (15), 16 - 22.

Rijksinstituut voor Volsgezondheit en Milieuhygiene 1986. ,,Criteriadocument Styreen."

Schmauder, H.-P., Frankenfeld, K., Ludwig, M. and Luthardt, W. 1997. ,,Purification of Contaminated Oil-binding Materials for Repeated Use." In B. C. Alleman and A. Leeson *In Situ and on-Site Bioremediation* Vol. 5, pp. 259-263, Battelle Press, Columbus, Richland

Tossavainen, A. 1978. ,,Styrene Use and Occupational Exposure in the Plastic Industry." *Scandinavian Journal of Work Environmental Health* 4, Suppl. 2, 7-13.

Vainio, H., Tursi, F. and Belvedere, G. 1992. ,,What are the Significant Toxic Metabolites of Styrene?" In: E. Hieten, M. Laitinen and O. Hanninen (Eds.), *Cytochrome P-450, Biochemistry, Biophysics and Environmental Implications,* pp. 678-687, Elsevier Biomedical Press B.V., Amsterdam

Acknowledgement: The authors are grateful for grants from the European Commission for this work (grants No. ERBIC 15CT96/0710 and ERBIC 15CT96/0716) and for additional financial aid from national governments.

ACCELERATING BIOREMEDIATION USING A PARTICULAR SURFACTANT

Bob Becker P.Eng. QEP
Colt Engineering Corporation Calgary, Alberta

INTRODUCTION

Soil can have non-aqueous phase liquid (NAPL) contaminate/s spilled onto or into it. When this happens the NAPL's tend to migrate down as a separate liquid phase and disperse into the soil matrix until they become trapped in the soil as immobile droplets or ganglia. NAPL's tend to follow high permeability path ways and spread along low permeability layers. The disposition of NAPL's in the soil matrix can be in one or all of the following forms:

- evaporate and stay in the vapor phase,
- adsorb onto solid surfaces
- dissolved into the soil moisture. (Wunderlich, R. W. et al):

In the case of a water table being in the migration route of the NAPL/s the path is determined by NAPL's

specific gravity(SG):

- Dense non aqueous-phase liquids (DNAPL's) have a SG greater than the groundwater and therefore continue on a downward migration and will spread out at the bottom of the aquifer.
- Light non aqueous-phase liquids(LNAPL's) have a SG less than the groundwater and will spread out on top of the groundwater.
- In both cases the NAPL will disperse into the soil matrix until some of contaminate becomes immobile droplets or ganglia. (Hunt, J. R. et al).

Usually NAPL's have a low solubility in water i.e. 5 ug/liter to 1000 mg/liter. However,regulations can require considerably lower concentrations thus requiring remediation. The slow release of the contaminate has often limited the effectiveness of many of the conventional soil treatment systems such as pump and treat, vacuum extraction and in-situ bioremediation. This is because the remediation process takes too long to achieve the desired limits.

POSSIBLE SOLUTION -USE SURFACTANTS

One of the techniques used to overcome the problem of the slow release of immobilized NAPL's is to solubilize them with surfactants. (Edwards, D. A. et al). Surfactants are capable of emulsifying LNAPL to facilitate increased mobility and recovery efficiency (Chevalier *et al.*, 1997; Abdul *et al.*, 1990). In many cases this technique can then enhance bioremediation if the surfactant is not toxic to the

NAPL degrading microorganisms. In most cases it has been found that surfactants do enhance bioremediation.

Surfactants are essential to the to the bioremediation process. This is borne out by the fact that microorganisms produce surfactants in order to solubilize hydrophobic organic compounds (Lange, S. and Wagner, F). Whether the surfactant is produced by microorganisms or manufactured they both act to solubilize the target compound. Surfactants can act in two ways : (1) increase solubility (solubilization) and (2) lower the interfacial tension (mobilization).

The lowering of the interfacial tension increases the mobility of the NAPL by reducing the capillary forces that immobilized it. Solubilization moves the NAPL from being adsorbed on the soil to the water phase as an emulsion. When solubilized the NAPL is then available to be metabolized by mico-organisims. Surfactant .field studies, primarily conducted at Hill Air Force Base in Utah, have attained up to a 99% source removal rate.

Toxicity and biodegradability of surfactants are also factors in bioremediation because they effect the efficiency of bioremediation system. There are a number of low toxicity and highly biodegradable surfactants. One such compound that we have had experience with is a patented product (BioSolve ^{TM}) that is also used as a fire suppressant. This product acts as surfactant in that it encapsulates (solubilises) the compound in water so that it can't vaporize. That is how it acts as a fire suppressant. It has both UL and ULC approval through testing which proved its capability to suppress the vapors of volatile organic compounds (VOC's).

After the compound is encapsulated there is an indication that BioSolve tends to disperse the compound more than most surfactants by making extremely small micelles in the aqueous phase. This dispersing action makes more of the compound available to the microorganisms. Some studies show the slowest step in biodegradation is accessing the carbon substrate (Churchill,S. A. et al). Thus the more the compound is dispersed in water the more it is available to the microorganisms. The main reason for adding a compatible manufactured surfactant to a bioremediation system is to speed up the process. It takes considerable time for micro-organisms to create enough surfactants accelerating the process. In some cases micro-organisms can not produce enough surfactant to start biodegradation

AQUEOUS DISPERSION OF OCTADACANE AND CO$_2$ CONVERSION USING TWO DIFFERENT SURFACTANTS

Tests of octadecane solubilization and carbon to CO$_2$ conversion tests at the University of Alabama using two commercially available surfactants showed: (1) BioSolve dispersed the octadecane 50% to 100% more at concentrations of 0.1 % to 0.5% than another popular surfactant and (2) After 68 hours BioSolve had a 48% conversion rate to CO$_2$ compared to 19% for bacteria only.

EFFECT OF A SURFACTANT ON HYDROCARBON DEGRADER MICROBIAL POPULATION AND ALKALINE DEGRADATION

Tests on BioSolve by the National Environmental Technology Applications Center (NETAC) at the University of Pittsburgh which included a series of runs comparing the microbiological enumeration of hydrocarbon degraders for the following samples:

- Control
- Control plus nutrient
- Control plus nutrient and BioSolve

These tests showed: (1) After 28 days the BioSolve Test had 4,500,000 organisms/ml where as the control and the nutrient only samples had less than 100,000 microorganisms/ml(2) After 28 days the control still had 25,000 PPM hydrocarbon whereas the nutrient only sample had 17,500ppm and the BioSolve sample had 7,500ppm

The runs show that the surfactant accelerates the microbial population and hydrocarbon degradation. It should be noted that the most common hydrocarbon degrader micro-organisms are aerobic (need oxygen and produce CO2 and water).

CO_2 PRODUCTION CONTROL TESTS WHEN USING A SURFACTANT

One way of assessing aerobic biological activity is to measure carbon to CO_2 conversion. The following five CO_2 conversion tests were conducted at the University of Alabama. A summary of the protocol used in five tests is as follows:

- For each test a quantity of thoroughly mixed soil or beach sand was divided into two equal parts
- Each part had an equal amount of radioactive carbon marked contaminant mixed into it.
- Each part had an equal amount of water mixed in, one part with only water and one part with water and surfactant
- Each part was enclosed and the CO_2 conversion was measured
- Each of the five tests had a different type of contaminate and/or different levels of contamination

The first test in which the contaminate was a relatively light refined hydrocarbon. After 11 days the CO_2 % conversion of the control was approximately 14.75 %. The surfactant reached the same % in 3.85 days. This indicates the surfactant speeded up the bioremediation process by close to a factor of 2.88.

Test #2 used crude oil as the contaminate. The control part of the test reached 23.5% CO_2 conversion after 90 hours where as the surfactant part reached the same level in about 17 hours. This implies the surfactant speeded up the process by over a factor of 5.

Tests #3 and #4 were with crude oil. The difference in these tests is how the oil contaminated the soil. Test #3 simulated a dry land situation i.e. pipeline break. The oil was poured on the soil and mixed then followed by the addition of water or water and surfactant then mixed again. This tends to create more oil wetted soil surfaces than when the soil is already water wetted. The control reached a CO_2 conversion of 8.33% after 120 hours, while the surfactant reached it in about 21.2 hours or a factor of approximately 6 times faster. Test #4 simulated a fresh water spill i.e. on a river. The soil was wetted with water and then oil was mixed with water before being put on the soil. The control reached a CO_2 % conversion of 30.5% in 90 hours. The surfactant reached the same % in 17 hours or a factor of over 5 times faster.

Test #5 simulated a spill onto a salt water beach. The soil had been preconditioned with marine micro organisms for seven. days. In this case the control indicated very little remedial action. The surfactant showed 45 % CO_2 conversion in 8 days indicating that a surfactant can start remediation that won't start on it's own.

COMPARISON OF A COMMONLY USED SURFACTANT WITH BIOSOLVE

A test which was part of a program to remediate sites of old crude oil spills where it appears the natural bioremdiation had virtually stopped. A series of tests conducted on contaminate that was a 5 year old heavy crude oil that had been weathered in a tropical environment. The tests consisted of four bioreactors set up with a specially developed culture of microorganisms, nutrients and aerated plus one reactor as control was setup without microorganisms, nutrients and aeration. The protocol was:

- Reactor # 1 had only the contaminated soil
- Reactor #2 had only microorganisms, nutrients and aeration;
- Reactor #3 had microorganisms, nutrients , aeration and 0.5% BioSolve.
- Reactor #4 had microorganisms, nutrients and aeration and 5.0% BioSolve.
- Reactor #5 had microorganisms, nutrients and aeration and a commonly used surfactant.

The results showed:

- Reactor # 1 had 0% removal indicating bioremediation of this crude had virtually ceased
- Reactor # 2 had 69 % removal indicating bioremediation can take place by using microorganisms, nutrients and aeration
- Reactor # 3 had 69 % removal indicating low concentration of the surfactant little or no effect
- Reactor # 4 had 95 % removal indicating with the 5.0% BioSolve solubilized the high molecular weight hydrocarbon and sped up the bioremediation leaving only 5% of the original contamination after 21 days, where as the other tests had 6 times more contaminate left..

- Reactor # 5 had 60 % removal indicating showed how the commonly used surfactant actually slowed the bio-remdiation process slightly. There have been cases were some surfactants significantly slow down bio-remediation (Falatko, D. M. ,et al.).

28 DAYS CO_2 PRODUCTION COMPARING CLEAN AND CONTAMINATED SOIL WITH AND WITHOUT A SURFACTANT

A control test was conducted at the University of California - Berkeley using an autoclave sterile gravel and fill soil matrices with CO_2 production being measured during a 28 day incubation period.. A summary of the protocol is as follows:

- Test # 1 was soil matrix and H_2O only,
- Test # 2 was soil, H_2O and BioSolve,
- Test # 3 was soil, H_2O and hydrocarbon contaminate,
- Test # 4 was soil, H_2O, hydrocarbon contaminate and 3% BioSolve.

These tests show the following:

- The soil matrix contained enough carbon for the micro-organism to produce 15 mg CO_2
- That BioSolve is biodegradable by producing 69mg CO_2
- The hydrocarbon produced 69mg CO_2 as food source by the micro-organisms
- BioSolve made more of the hydrocarbon available as a supplementary food source by producing 175mg of CO_2

A CASE STUDY OF IN-SITU BIOREMEDIATION USING A SURFACTANT

The site on which this in-situ bioremediation took place was in Bakersfield California for an agricultural manufacturer . The contamination was caused by two 2270 liter leaking underground storage tanks. The contamination went to a depth of 13.7m. The plot plan area of the contamination was 10m by 10m. Thirteen monitoring/inoculation wells were drilled. Nine on an outside circumference of plume and four in the plume. These were used to measure VOC levels and in-oculate the surfactant (BioSolve), nutrients, air(by vacuum) and microorganisms. Three sample spots on the periphery of the plume and two sample spots in the center of the plume were used to sample the soil. The following summarizes the high, low and average VOC readings from the wells on this project.. An inter-esting occurrence was observed in the results is the increase of the high VOC val-ues 12 days after the remediation process began. This is caused by some of the previously immobilized hydrocarbon which couldn't be detected were mobilized.

EVENT DATE	DESCRIPTION of ACTIVITY	VOC LOW	in AVE	ppm HIGH
8/6/93	Wells installed and intial reading	210	4100	9600
8/7/93	Initial nutrient and water addition on new wells	200	3835	8750
8/9/93	Second nutrient treatment. Vapor recovery installed	200	2925	8500
8/17/93	2% BioSolvetm to all but one well OVA/PID VOC readings	9	1230	10000
8/18/93	Inoculation of microorganisms, nutrients and BioSolve			
8/20/93	OVA/PID readings for VOC's	10	119	1000
9/10/93	Inoculation of nutrients and BioSolve			
9/22/93	Inoculation of nutrients and BioSolve OVA/PID readings	18	177	1500
10/1/93	Inoculation of nutrients and BioSolve			
10/6/93	Inoculation of nutrients and BioSolve OVA/PID readings	10	45	350
10/13/93	OVA/PID readings for VOC's	15	54	104
10/21/93	OVA/PID readings for VOC's	14	49	200
10/28/93	OVA/PID readings Inoculation of nutrients and BioSolve	12	31	110
11/15/93	OVA/PID readings for VOC's	10	36	120
11/30/93	OVA/PID readings for VOC's	10	31	86
12/07/93	OVA/PID readings Inoculation of nutrients and BioSolve	10	38	100
12/22/93	OVA/PID readings for VOC's and soil sample	10	26	105
1/10/94	Inoculation of nutrients and BioSolve			
1/20/94	OVA/PID readings Inoculation of nutrients and BioSolve	5	22	80
2/10/94	OVA/PID readings for VOC's	8	16	50
2/25/94	OVA/PID readings for VOC's	5	12	28
3/01/94	**SOIL TESTS WERE BELOW DETECTABLE LIMIT**	.005	.005	.005
5/15/94	SITE CLOSED			

Unfortunately in this situation a control could not be used. However this was an old spill and the initial levels give an indication of what a control would look like. This protocol was used because it was estimated to be 38%-45% less expensive than thermal treatment and more acceptable to the public. Other benefits of this process are: (1) The bioremediation continues on after the site closing which can result in the ultimate destruction of the hydrocarbons; (2) The surfactant minimizes the VOC's from migrating from the ground.

A CASE STUDY USING A SURFACTANT TO FLUSH HYDROCARBONS FROM THE SOIL

A SEAR (surfactant enhanced aquifer remediation) pilot project was conducted by Groundwater & Environmental Services, Inc. (GES) of Wall, New Jersey to evaluate *in-situ* surfactant flushing as a viable option for source area remediation at a service station site. The study area included the following conditions:

- The subsurface stratigraphy consists of silt with little fine sand and clay.
- Depth to groundwater is approximately 4 to 5 feet below grade.
- The average hydraulic gradient and hydraulic conductivity are 0.04 feet/foot and 3.45 x 10^{-5} centimeters/second, respectively.
- LNAPL was present at one monitoring well

- The average hydraulic gradient and hydraulic conductivity are 0.04 feet/foot and 3.45 x 10^{-5} centimeters/second, respectively.
- LNAPL was present at one monitoring well
- The area of hydrocarbon-impact (LNAPL and residual LNAPL) was estimated at approximately 300 f t^2, averaging approximately 6 to 8 inches in thickness.

The actions taken are as follows:

- Between May of 1993 and April of 1997 LNAPL thickness in the impacted well ranged from a sheen to 0.30 feet due to the fluctuating water table. Manual bailing resulted minimal LNAPL recovery. Continual occurrence and minimal thickness of LNAPL indicated capillary entertainment within the "smear zone"
- In June of 1997 GES proposed SEAR pilot study to the State of New Jersey Department of Environmental Protection that included the injection of 250 gallons of 2% (v/v) BioSolve into three monitoring wells
- In June of 1997 a baseline groundwater analysis of Methyl Blue Active Substances (MBAS none found) was conducted.
- July of 1997, LNAPL monitoring immediately prior to the surfactant injection detected a thickness of 0.28 feet.
- July of 1997, 155gal. of a 2% solution of BioSolve® was injected via gravity/siphon feed into the impacted well and the two closest monitoring wells
- After 24 hours the total phase (vapor and liquid) from the LNAPL-impacted well and a second surfactant injection well was recovered.
- After another 3.5 hours, the LNAPL-impacted well was placed off-line and recovery on the third injection well was started.
- An additional 20 gallons of 2% BioSolve® solution were injected into the LNAPL-impacted well.
- After an additional 1.5 hours of recovery, one well was taken off-line and the LNAPL-impacted well was placed back on-line for the remainder of the test (additional 1 hour).

During the recovery phase the following was observed:

- Influent concentrations from the LNAPL-impacted well ranged from 5,550 to 50,000 parts per million volume (ppm$_v$).
- Influent concentrations from the other injection points varied between 50 and 4,000 ppm$_v$ and 22 and 33 ppm$_v$, respectively.
- The calculated hydrocarbon removal rates from the LNAPL-impacted well ranged from 0.94 to 9.90 pounds/hour.
- The calculated removal rates from the other injection wells varied, one ranged from 0.01 to 0.56 pounds/hour while the other remained at 0.01 pounds/hour.
- The cumulative mass of hydrocarbons removed from the vapor phase was calculated to be 35.7 pounds.
- In addition, 255 gallons of hydrocarbon-impacted groundwater and emulsified LNAPL were removed during the recovery phase. As the LNAPL was recovered in an emulsified aqueous phase, the volume of LNAPL recovered could not be accurately measured.

- The laboratory analysis of post-injection groundwater samples indicated MBAS concentrations of 0.11 mg/L and 0.09 mg/L.

Conclusion of pilot project: groundwater monitoring conducted to date since the SEAR pilot test indicates that no LNAPL is present at any well on-site. It was concluded that the SEAR pilot study effectively mitigated the presence of LNAPL in the area of the former fuel dispenser island. A workplan proposing natural attenuation as the remedial action on-site was submitted to the NJDEP and is currently under

CONCLUSION

From the bench and field scale tests it appears that bioremediation can be speeded up through the use of surfactants.

There are some cases where bioremediation can only occur through the application of a surfactants. This was demonstrated in the control CO_2 production tests on beach sand

The field test showed that surfactant application can be used in in-situ situations. Also in-situ solubilization and mobilization of NAPL's through the use of surfactants is becoming better documented. (Fountian, J. C.,and C. Waddell-Sheets)), (Fountain J. C.), Wayt, H. J. and D. J. Wilson) ,(Vigon, B. W. and A. J. Rubin).

REFERENCES

Abdul, A.S., T.L. Gibson, and D.N. Rai. 1990. Selection of Surfactants for the Removal of Petroleum Products from Shallow Sandy Aquifers. Ground Water. Vol. 28, No. 6. p920-926.

Churchill, S. A.; Jefcoat, I.A. and Churchill, P. F. (1994) "Anionic Surfactant Enhanced Hydrocarbon Degradation by Pure Stains and Bran Sorbed Micrbial Consotia", University of Alabama

Edwards, D. A. et al (1991) Surfactant Enhanced Solubility of Hydrophobic Organic Compounds in water and in Soil-Water systems. In Organic Substances and sediments in Water. Vol. 2, R.A. Baker Editor, Lewis Publishers, Chelsey MI.

Falatko, D. M. ,et al Effects of Biologically Produced Surfactants on the Mobility and Biodegration of Petroleum Hydrocarbon, Vol. 6, No. 2, Water Environment Research, Washington D.C.

Fountain J. C. "REMOVAL OF NON-AQUEOUS PHASE LIQUIDS USING SURFACTANTS", State University of NY Buffalo.

Fountain, J. C.,and C. Waddell-Sheets, "A Pilot Scale Test of Surfactant Enhanced Pump and Treat" State University of NY Buffalo

Hunt, J. R. et al 1988 Nonaqueous phase liquid transport and cleanup. 1. Analysis of mechanism. Water Resour. Res.. 24(8), 1247-1258

Lange, S. and Wagner, F. (1987) Structure and Properties of Biosurfactants. In Biosurfactants and Biotechnology . Marcel Dekker Inc., NY

Lenhart,R. 1997, Summary Report "NCP Product Schedul Bioremediation Agent Efficacy Evaluation of BioSolve".

Melrose, J. C. and Brander, C. F. 1974. Role of capillary forces in determining microscopic displacement efficiency for oil recovery by waterflooding. J. Can. Petroleum Technol. 54-62).

Moore, J. W. (1989)

Vigon, B. W. and A. J. Rubin 1989. Practical Considerations in the Surfactant-aided Mobilization of Contaminates in Aquifers. J. Water Pollution control Federation, 61:1233-1240.)

Wayt, H. J. and D. J. Wilson Soil Clean UP by In-situ Surfactant Flushing. Marcel Dekker, Inc., Separation Science & Technology, 24: 905-937

Wunderlich, R. W. et al . 1992. In Situ Remdiation of Aquifers Contaminated with Dense Nonaqueous Phase Liquids by Chemically Enhanced Solubilization J. Soil Contamination, 1(4):361-378.

EFFECTS OF SURFACTANTS ON HYDROCARBON BIODEGRADATION IN LANDFARMING

Claudio O. Belloso (Facultad Católica de Química e Ingeniería, Rosario, Argentina)

ABSTRACT: A pilot-scale test was performed with the purpose of determining the effect produced by a surfactant ionic, Novel II, on hydrocarbon biodegradation in landfarming soils inoculated with commercial and indigenous bacterial strains. Inoculated soils were maintained outdoors for 210 days.

The presence of the surfactant Novell II seems to reduce the biodegradation of the tested commercial strains. The rate of biodegradation obtained with the W2 indigenous strain is greater when the soil contains surfactant.

Application of surfactants must be done with caution if commercial or indigenous bacteria are inoculated.

INTRODUCTION

Bioavailability of low-solubility hydrocarbons in soils is limited by their sorption onto the soil matrix. Although the capacity of soils to detoxify waste has been well documented, this capacity is limited and natural detoxification processes often require years to restore impacted sites.

Sorption and sequestration of hydrocarbons within the soil matrix are critical processes affecting contaminant mobility, toxicity, and persistence. Slow desorption and release from the soil matrix to the aqueous phase represents a long-term contaminant source and hinders remediation efforts.

Surfactants above their critical micelle concentration enhance desorption of these hydrocarbons, and the fraction of hydrocarbon dissolved in the micelles can be bioavailable depending on the surfactant structure and its concentration in the aqueous phase.

Both synthetic and biological surfactants have been shown to enhance the apparent aqueous solubility of nonpolar organic compounds resulting in increased bioavailability and biodegradation. However, there are also reports which suggest that some synthetic surfactants inhibit biodegradation. This inhibition is generally attributed to toxicity or reductions in bioavailability due to partitioning of the contaminant into surfactant micelles.

The surfactant behavior in soils is complex, since surfactants may solubilize the pollutants, but may also sorb onto the soil, providing an additional phase that increases sorption, and may also have an oxygen demand. Therefore, even though surfactant-enhanced bioremediation of contaminated soils is a promising technique, it will only work if surfactant and their dose are selected carefully with consideration of site-specific conditions (Guha and Jaffi, 1996).

Studies demonstrated that non-ionic surfactant Novel II notably improved biodegradation in contaminated soils with phenanthrene and biphenyl (Aronstein and Alexander, 1993).

Objective. To determine the effect produced by a non-ionic surfactant, Novell II, on hydrocarbon biodegradation in landfarming soils inoculated with commercial and indigenous bacterial strains.

MATERIALS AND METHODS

Eleven metallic containers with a 50 L capacity were used. Thirty kg of soil per container was added. Soil was obtained from a landfarming site belonging to the Petroleum Refinery San Lorenzo S. A. located in San Lorenzo City, Santa Fe Province, Argentina. Table 1 illustrates the percent composition of hydrocarbon residual in the landfarming soil. The initial Total Organic Carbon content of the landfarming soil was 1.56% w/w.

To avoid contact of the soil with the metal of the container, the containers' inner walls were covered with a thick layer of non-biodegradable resin.

TABLE 1. Percent composition of hydrocarbon residual contained in landfarming soil

Hydrocarbon distribution	Percentage
Saturates	24
Aromatics	33
Resins	17
Asphaltenes	26

Containers N° 3, 4 and 5 were inoculated with a hydrocarbon degrading bacteria denominated W2 isolated from the same landfarming site (Belloso et al., 1997). Containers N° 6, 7 y 8 were inoculated with a D-5000 commercial strain and containers N° 9, 10 and 11 were inoculated with a D-5075 commercial strain.

One application of NPK type fertilizer was made at the beginning of the trial. Fertilizers were formulated and added according to the proportions shown in Table 2.

TABLE 2. Fertilizers application and formulations

Denomination	Relationship C/N	Relationship NPK	Source of nitrogen
Z2	100	20:20:1	NH_4^+
Z21	200	20:20:1	NH_4^+

In containers N° 3, 4, 6, 7, 9 and 10, fertilizer Z2 was added. In containers N° 5, 8 and 11, fertilizer Z21 was added. In soils contained in containers N° 4, 5, 7, 8, 10 and 11, 1,000 mL of an aqueous solution of Novel II (0.1 g/L) per container was added. Container N°1 contained only soil as a control. Container N°2 contained soil and $HgCl_2$ (2g/100g of dry soil) in order to evaluate abiotic losses of hydrocarbons (Ferrari et al., 1994).

All containers were maintained outdoors for 210 days, from July 1997 to February 1998 once inoculated. Soils were aired weekly by blending the whole content.

At the beginning of the trial and every 30 days the following parameters in each soil were determined: pH, moisture, total hydrocarbons, total aerobic heterotrophic bacteria (TAHB) and hydrocarbons degrading microorganisms (HDM).

The pH was measured in suspensions of soil (40% w/v) in $CaCl_2$ 0.01M solution. Moisture in soil samples was determined by drying to 105°C until a constant weight was reached. The hydrocarbon content was determined by extraction in Soxhlet with Ethylic Ether for 6 hours. TAHB in soils was enumerated by the pour plate method (Clesceri et al., 1992) using Merck plate count agar. All plates were incubated at 30°C for 24 hours. The results of TAHB count at the beginning of the trial are shown in Table 3.

HDM in soils was enumerated by the more probable number method (Clesceri et al., 1992) using Bushnell-Hass mineral broth (Bushnell and Hass, 1941) with n-hexadecane as the sole carbon source. The results of HDM count at the beginning of the trial are shown in Table 4.

A five-tube MPN technique was used. Three sets of five screw-cap tubes, each containing 5 mL of Bushnell-Hass mineral broth were inoculated with 1, 0.1, and 0.01 mL of a sample dilution. After inoculation, 50 µL of UV-sterilized n-hexadecane (Merck) and 100 µL of autoclaved resazurin solution (Merck, 50 mg/L) were added to each tube. After incubation at 30°C for 7 days, tubes giving positive reaction were counted. The color of positive tubes changed from blue to pink.

RESULTS AND DISCUSION

In 210 days, the moisture content varied between 14 and 17%. The pH varied between 6.3 and 7.0. The average temperature was 20° C.

As we can see in Tables 3 and 4, there were not very remarkable changes in TAHB count (average value for 210 days: 1.2×10^7 CFU/g) and HDM counts (average value for 210 days: 9.1×10^7 MPN/g).

In Table 5, we can see that the biggest percentage of HC reduction was obtained in container 6.

CONCLUSIONS

The presence of surfactant Novell II seems to reduce the biodegradation of the tested commercial strains. The rate of biodegradation obtained with the W2 strain is greater when the soil contains the surfactant; however, the poor biodegradation detected in general in the containers in which this strain was inoculated was unexpected. In previous tests (Belloso et al., 1997), this same strain had demonstrated to possess a great capacity for biodegradation. Possibly, successive transfers in culture media have modified this property.

The presence of surfactant with commercial strain D-5075 has influenced biodegradation negatively. Application of surfactants must be accomplished with caution if commercial or indigenous bacteria are inoculated.

TABLE 3. TAHB count in each container during test

Container N°	CFU/g	CFU/g	CFU/g	CFU/g
1	1.4×10^7	4.8×10^6	3.6×10^7	2.3×10^7
2	0	0	0	0
3	1.5×10^7	1.7×10^7	5.7×10^7	3.7×10^7
4	5.3×10^7	7.2×10^6	6.4×10^7	3.6×10^7
5	6.3×10^7	1.1×10^7	5.9×10^7	3.4×10^7
6	2.0×10^7	9.7×10^6	9.9×10^7	7.0×10^7
7	1.7×10^7	3.5×10^6	6.6×10^7	3.5×10^8
8	1.9×10^7	2.9×10^6	4.7×10^7	1.6×10^7
9	1.5×10^7	4.8×10^6	3.5×10^7	1.9×10^7
10	3.5×10^7	1.2×10^6	4.0×10^7	2.1×10^7
11	2.7×10^7	4.1×10^6	2.8×10^7	1.1×10^7

TABLE 4. HDM count in each container during test

Container N°	MPN/g	MPN/g	MPN/g	MPN/g
1	5.4×10^7	2.9×10^7	1.3×10^8	1.3×10^8
2	0	0	0	0
3	1.3×10^8	5.5×10^7	1.3×10^8	1.8×10^7
4	$>1.3 \times 10^8$	2.5×10^7	5.4×10^7	1.3×10^8
5	5.4×10^7	1.3×10^8	5.4×10^7	1.3×10^8
6	5.4×10^7	1.4×10^7	1.3×10^8	5.5×10^7
7	1.1×10^7	5.4×10^7	1.3×10^8	5.5×10^7
8	1.3×10^8	2.5×10^7	1.3×10^8	1.8×10^7
9	5.3×10^8	1.3×10^8	5.4×10^7	1.3×10^8
10	5.3×10^7	2.8×10^7	5.4×10^7	1.8×10^7
11	1.1×10^7	1.3×10^8	2.8×10^7	2.5×10^7

TABLE 5. Percentages of HC reduction

Container N°	Strain, surfactant and fertilizer added	HC (% w/w) 0 day	HC (% w/w) 210 days	HC (% w/w) reduction obtained by linear regression
1	Control	5.4	4.0	16
2	Hg	4.7	4.8	26
3	W2-Z2	3.1	5.7	-76
4	W2-NOVEL II-Z2	4.6	3.0	-27
5	W2-NOVEL II-Z21	6.0	3.4	21
6	D-5000-Z2	4.1	2.5	53
7	D-5000-NOVEL II-Z2	3.3	2.0	41
8	D-5000-NOVEL II-Z21	3.2	2.1	34
9	D-5075-Z2	5.2	2.7	45
10	D-5075-NOVEL II-Z2	5.2	5.1	-15
11	D-5075-NOVEL II-Z21	3.7	3.6	-5

ACKNOWLEDGEMENTS

The author thanks the authorities of the Faculty and Refinery San Lorenzo that believed in the potential of this project and supported it.

REFERENCES

Aronstein, B. N., and M. Alexander. 1993. "Effect of a non-ionic surfactant added to the soil surface on the biodegradation of aromatic hydrocarbons within the soil." *Applied Microbiology and Biotechnology.* 39(3): 386-390.

Belloso, C. O., J. Carrario, and D. Viduzzi. 1997. "Hydrocarbons degrading-bacteria isolated of contaminated soils with hydrocarbons" (in Spanish). In *Proceeding III Jornadas Rioplatenses de Microbiología.* Asociación Argentina de Microbiología. pp. 76, E9. Buenos Aires, Argentina.

Bushnell, L. D., and H. F. Hass. 1941. "The utilization of certain hydrocarbons by microorganisms". *J. Bacteriol. 41*: 653-673.

Clesceri, L. S., A. E. Greenberg, and R. Trussel Rhodes (Eds.). 1992. "Standards Methods for the Examination of Water and Wastewater". 18th ed., American Public Health Association, N.Y.

Ferrari, M. D., C. Albornoz, and E. Neirotti. 1994. "Biodegradability on soil of residual hydrocarbons from petroleum tank bottom sludges" (in Spanish). *Revista Argentina de Microbiología, 26*: 157-170.

Guha, S., and P. R. Jaffi. 1996. "The Bioavailability of Hydrophobic Compounds Partitioned into the Micellar Phase of Nonionic Surfactants." *Environmental Science and Technology 30*(4): 1382-1391.

STABLE CARBON ISOTOPE ANALYSIS: A TOOL TO MONITOR BIOREMEDIATION STATUS AND ACHIEVE SITE CLOSURE

H. James Reisinger, Aaron S. Hullman, John M. Godfrey (Integrated Science & Technology, Inc., Marietta, Georgia, USA)
Daniele Arlotti (Foster Wheeler Environmental Italia S.r.l. Milan, Italy)
Giorgio Andreotti and T. Ricchiuto (ENI S.p.A. – AGIP Div. Milan, Italy)
Robert E. Hinchee (Parsons Engineering Science, South Jordan, Utah, USA)

ABSTRACT: Bioremediation, including bioventing, biopiling, and landfarming, was used to remediate deep and surficial soil, respectively, impacted with crude oil as the result of an oil well blowout in Trecate, northern Italy in 1994. The bioventing system and the biopiles were engineered and key process control parameters were strictly controlled to achieve maximum mass removal efficiency and the very low numerical endpoints established by Italian regulatory authorities at the outset of the project. Further, the bioventing system and the biopiles were monitored using both *in situ* respirometry and mass removal calculations and plotting. After approximately 12 months of operation, the mass removal plots showed that mass removal in the biopiles had achieved its asymptotes -- an indication of biodegradation rate decline and a strong suggestion of substrate limitation. However, *in situ* respirometry test results showed that the microbial community continued to mineralize carbon of unknown character at a significant rate, which suggested that the biodegradation process was ongoing. Stable carbon isotope analysis of the expelled crude oil, contaminated soil, background soil, wood chips, soil vapor, and ambient air was used to determine whether the carbon being oxidized by the microbial consortium was from naturally occurring material in the soil (i.e., humus) and wood chips used to bulk the biopile soils or from residual crude oil. This discrimination, along with the asymptotic mass removal plots, served as the basis to conclude that bioremediation had proceeded to its logical endpoint, and that continued biopile operation would not appreciably reduce the hydrocarbon mass.

INTRODUCTION

While bioremediation is a valuable environmental cleanup tool, and one that is becoming more commonly applied and readily accepted, it is often difficult to document progress and determine when its logical endpoint has been achieved. All too often, numerical endpoints are established at the outset of a bioremediation project based not on the limitations of the kinetics or dynamics of the site-specific biological processes, but rather on a predetermined regulatory limit. This approach often leads to the conclusion that the bioremediation process has failed completely, that it has not operated efficiently, or that it was improperly applied. In most instances, the problem rests in a lack of understanding of the concepts of substrate limitation and population carrying capacity which, together, work to dictate the achievable concentration endpoint. Many times, standard bioremediation monitoring approaches fail to provide the information needed to determine whether progress is being made and when the bioremediation process has reached its logical endpoint. Field respirometry, a commonly applied approach, is particularly problematic from this perspective, as it is a measure of total community aerobic activity regardless of substrate. That is, it is a manifestation of the degradation of both target compounds or analytes as well as non-target, naturally occurring organic material.

Stable carbon isotope analysis is a tool capable of discriminating between the carbon sources being used by a microbial consortium in a biologically based remediation system. The approach is based on the fact that the carbon that comprises all living things is made up of several stable isotopes, most notably ^{13}C and ^{12}C. ^{12}C and ^{13}C generally occur in nature at a ratio of approximately 98.89percent ^{12}C to 1.11percent ^{13}C. These stable carbon isotopes are present in all living organisms and in materials derived from living organisms at a ratio ($\delta^{13}C$) unique to them. The ratio can be changed as a result of natural processes such as hydrocarbon mineralization, and this phenomenon is termed fractionation. It is this unique ratio that provides for discrimination of carbon sources and renders the approach a valid and valuable bioremediation assessment tool.

Stable carbon isotope analysis was used in concert with *in situ* respirometry testing, mass removal analysis, and statistical analysis to evaluate progress and determine the point of completion (i.e., process-based rather than numerically based) for a large crude oil bioremediation project in Italy.

Site Description. In February of 1994, in the course of drilling an oil exploration well in Trecate in northern Italy (Trecate 24), a high pressure zone was encountered which resulted in a blowout. Approximately 15,000-m^3 of light sweet crude oil was expelled from the well and deposited over approximately 5-km^2 of agricultural lands (i.e., rice and maize) downwind of the well. Following initial free oil recovery, it was determined by the Italian authorities that because the soils had unique pedological characteristics, remediation would need to be by a method that left them unchanged. Bioremediation was selected as the cleanup tool.

Soils were partitioned by concentration. Those with hydrocarbon concentrations in excess of 10,000-mg/kg were excavated and treated in two biopiles. The area excavated was 12.5-ha and the total excavated soil volume was approximately 26,000-m^3. Residual hydrocarbon in the deeper subsurface soils beneath the excavated 12.5-ha was treated via bioventing, an *in situ* bioremediation tool. The remainder of the impacted surficial soils were treated through large scale landfarming.

The biopiles were constructed in horizontal lifts and vertical zones and were outfitted with air injection systems, a water and nutrient delivery system, a heat trace, a leachate collection system, and a plastic cover system. The biopiles were 150-m by 50-m by 3-m. Prior to installation in the biopiles, the soils were dried to an average moisture content of approximately 15 percent from original moisture content as high as 30 percent) and they were bulked with 20 percent poplar wood chips to enhance air flow characteristics. The average residual hydrocarbon concentration in the biopiles was approximately 15,000-mg/kg.

The biopiles were monitored daily to ensure that key operational parameters (i.e., oxygen, temperature, moisture content) were held at optimal. *In situ* respiration tests were run monthly to evaluate the status of biological activity and to generate the data needed to estimate respirometrically derived mass removal. These data were used, in concert with the daily process control monitoring data, to optimize the systems. On a quarterly basis, soil samples were collected from the biopiles in which the concentrations of total petroleum hydrocarbons (TPH) and selected polynuclear aromatic hydrocarbons (PAH) were determined. These data were, in turn used to estimate hydrocarbon mass removal which was tracked and plotted. After about one year of operation, the biopile hydrocarbon mass removal data achieved their statistically derived asymptotes. That is, hydrocarbon mass removal had slowed to a point that it was apparently insignificant ($p < 0.05$). However, results of the respiration testing showed that the respiration rates were still significant (some in excess of 40-mg/kg·day). Thus, the analytically derived and the respirometrically derived hydrocarbon mass removal data were contradictory. It was hypothesized that the carbon being oxidized by the biopile consortia were from the humus in the agricultural soils or from the wood chips used to bulk the soils. Stable carbon isotope analysis was used to solve this conundrum.

The deeper impacted soils (i.e., from surface to the water table) occupied an area of about 12.5-ha and were on average about 7-m thick. Thus, the volume of impacted deeper soils was 8,750,000-m^3. This unit, which was comprised of coarse sand and gravel with cobbles, was treated through bioventing. The system included 26 air injection points which were served by five blower stations. Air was provided by rotary lobe compressors. Biologically based hydrocarbon mass removal was monitored via *in situ* respirometry through a network of tri-level monitoring points. Respiration tests were run quarterly and the data were used to evaluate progress. Soil vapor samples were also collected from the biovented area and analyzed for stable carbon isotopes.

MATERIALS AND METHODS

Stable carbon isotope analyses were carried in the ENI S.p.A. – AGIP Division laboratories in Milan, Italy. The analyses were done using mass spectrometry with gas chromatographic preparation. In the method used, the carbon examined was in the form of carbon dioxide which was separated from the gaseous biopile and bioventing system samples and collected from the products of thermal oxidation of the solid samples (i.e.,, Trecate crude oil, poplar chips, and soil). Stable carbon isotope ratios were generated through the mass spectrometric analysis and the results were normalized to the PDB (Pee Dee belemnite) standard. This standard, introduced by Craig in 1957 (Clark and Fritz 1997) is based on the internal calcite structure from fossil *Belemnitella americana* found in the Cretaceous Pee Dee Formation in South Carolina, USA. The isotopic ratio values, relative to the PDB standard are reported in units of parts per thousand (per mil) according to the following expression:

$$\delta^{13}C_{PDB}(permil) = \frac{(R_{sample} - R_{s\,tan\,dard})}{R_{s\,tan\,dard}} x10^3$$

Where $R = \delta^{13}C$. Thus, the values produced through the analysis are either positive or negative values relative to the PDB standard (Suchomel et. al. 1990).

At the Trecate site, in order to achieve the degree of biodegradation source discrimination desired, samples of the various potentially biodegradable materials included in the biopiles were collected and analyzed as described above. These included the crude oil that was expelled during the blowout, unimpacted agricultural soil of the same character as the soil impacted by the blowout, poplar chips used to bulk the soils in preparation for biopile construction, and ambient air. With the exception of the ambient air sample, the reference samples were prepared for mass spectrometer analysis by converting the carbon in the materials to carbon dioxide through thermal oxidation. The carbon dioxide was then separated and analyzed.

With the reference data generated, a biopile and bioventing monitoring program was initiated. Soil vapor samples were collected from selected locations within the biopiles and from within the biovented area via aspiration into evacuated 22-ml glass vials. Locations were concentrated in those areas where the carbon dioxide concentrations were highest which were indicative of higher levels of aerobic biological activity. Following collection, the samples were transported to the AGIP laboratory where they were analyzed as described above.

RESULTS AND DISCUSSION

Results of analysis of the reference materials including the crude oil that was expelled during the blowout, unimpacted agricultural soil of the same character as the soil impacted by the blowout, poplar chips used to bulk the soils in preparation for biopile construction, and ambient air are given in Table 1.

TABLE 1. Trecate Bioremediation Project
Stable Carbon Isotope Analysis Results

Sample	Sample Date	Mean δ^{13} C CO_2	Max δ^{13} C CO_2	Min δ^{13} C CO_2	Standard Deviation
Trecate Crude Oil	1994	-30.3	NA	NA	NA
Trecate Crude Oil	1997	-30.4	NA	NA	NA
Poplar Chips	1997	-27.4	NA	NA	NA
Ambient Air	1997	-7.5	-7	-8	NA
Background Soil	1997	-18.08	-14.6	-22.7	2.904

Examination of these data shows a relatively broad range of δ^{13}C CO_2 values (i.e.,, from the heavier or more positive values for ambient air to the lighter more negative values for Trecate Crude oil). Clearly, the values generated from the Trecate crude oil samples were lower or lighter (i.e.,, more negative) than those from all other materials. Furthermore, the spread observed in the values from material to material is also large. For example, the smallest spread is about three parts per mil between Trecate oil and poplar chips. This relatively large spread renders the tool more discriminating for this set of samples. That is, the data spread provides for more positive determination of the source of the carbon that produced the carbon dioxide. The spread also the tool more useful in determining the source of the carbon that produced the carbon dioxide in the biopile and bioventing soil vapor samples. The spread of values between reference sample types is further validated by the very high level of reproducibility of the analytical method. Aggarwal and Hinchee (1991) reported that for samples with approximately 0.1 percent carbon dioxide, the method reproducibility is approximately plus or minus one part per thousand. All samples in this study contained appreciably more carbon dioxide than this.

Results of analysis of soil vapor samples collected from the Trecate biopiles and the bioventing system are given in Table 2.

TABLE 2. Trecate Bioremediation Project Stable Carbon Isotope Analysis Results

Sample	Sample Date	Mean δ^{13} C CO_2	Max δ^{13} C CO_2	Min δ^{13} C CO_2	Standard Deviation
Biopile 1	2-Apr-97	-28.051	-24.010	-29.500	0.943
	16-May-97	-28.234	-25.160	-29.710	0.768
Biopile L	10-Apr-97	-28.011	-14.650	-30.610	2.653
	23-May-97	-27.386	-15.190	-30.000	3.002
Bioventing	8/11-Jul-97	-30.150	3.900	-52.200	7.451
	Dec-97	-27.647	-1.930	-35.090	5.572

Examination of the data shown in Table 2 shows that the mean δ^{13}C CO_2 values generated from the biopile soil vapor samples were clearly more positive than those from the Trecate crude oil (Table 1). The difference is generally in excess of two to three parts per mil which is well outside the level of analytical reproducibility. Comparison of the mean biopile values with the other reference materials shows that the biopile soil vapor sample values are, in general, closer to those of the wood chips. The biopile values are considerably more negative than those of the soil. Certainly, the δ^{13}C CO_2 ranges show that there are some values among the samples analyzed nearer to those associated with the Trecate crude oil. This variability is a reflection of the heterogeneity of the soil that was used to construct the biopiles. These results show that the biopile δ^{13}C CO_2 values is most indicative of wood chip biodegradation. The data suggest that there was little crude oil degradation occurring at the time of sample collection. The values also show no appreciable contribution from the humic material in the unimpacted soils. There also appears to have been little if any, atmospheric air contribution given the negativity of the biopile values.

The results of the biopile stable carbon isotope analysis were consistent with field observations and the asymptote analyses carried out on the biopile data. The data suggested, based on mass removal rate trends, that little crude oil was being biodegraded. Results of the stable carbon isotope analysis supported this observation. Coupled with the mass removal rate data, the stable carbon isotope data showed that the biological consortia in the biopiles had shifted from petroleum degrading heterotroph dominance to organisms that apparently favored carbon in the wood chips used to bulk the soils. This would also seem to suggest that most of the crude oil constituents that could be mobilized and could be used by microorganisms as a carbon source had been removed and consumed resulting in a relatively recalcitrant residual hydrocarbon mix. This was born out by comparative gas chromatographic analysis which showed that the crude oil composition had indeed shifted. Using a combination of mass removal rate data, statistical analysis, and stable carbon isotope data, a demonstration was prepared to document that the biopile remediation process had proceeded to its logical endpoint. The demonstration was accepted by the regulatory authorities and the biopiles were dismantled and the soil was returned to the fields from which it had been excavated to be put back into agricultural production.

Examination of the bioventing data (Table 2) shows that the $\delta^{13}C$ CO_2 continues to be more negative and closer to that of the Trecate crude oil. This is not surprising, because, despite considerable documented biodegradation through bioventing, some of the subsurface continues to contain phase separated crude oil on the water table. Given that the water table fluctuates broadly as a result of large scale flood-type irrigation, the vadose zone is re-coated with oil annually. In this case, the stable carbon isotope data show that bioventing is continuing to remove crude oil through biodegradation.

In summary, stable carbon isotope analysis has proven to be a valuable and defensible tool applicable to the management of bioremediation projects. Its utility has been shown in both *in situ* and *ex situ* applications. Coupled with more conventional mass removal trend analysis, stable carbon isotope analysis can be used to demonstrate completion of biodegradation-based processes and to achieve site closure.

REFERENCES

Aggarwal, P.K. and R.E. Hinchee. 1991. "Monitoring In Situ Biodegradation of Hydrocarbons by Using Stable Carbon Isotopes." *Environ. Sci. Technol.* 25(6): 1178-1180.

Clark, I. And P. Fritz. 1997. *Environmental Isotopes in Hydrogeology*, pp. 1-34. Lewis Publishers, Boca Raton, FL.

Suchomel, K.H., D.K. Dreamer, and A. Long. 1990. "Production and Transport of Carbon dioxide in a Contaminated Vadose Zone: A Stable and Radioactive Carbon Isotope Study." *Environ. Sci. Technol.* 24(12): 1824-1831.

COMPOSTING CRUDE OIL-IMPACTED SOIL: PERFORMANCE COMPARISON WITH LAND TREATMENT AND SOIL PRODUCTIVITY IMPLICATIONS

Rob Hoffmann (Chevron Canada Resources, Calgary, Alberta)
Donna Chaw (Olds College Composting Technology Centre, Olds, Alberta)

ABSTRACT: A composting demonstration in a full-size windrow was conducted in 1997 using crude oil-affected soil to assess whether composting enhanced the extent of total petroleum hydrocarbon (TPH) removal relative to land treatment, and whether composting restored vegetative capability to pre-impact levels. TPH and n-alkane data suggest that composting and land treatment were equally effective in removing crude oil constituents, though composting enhanced the extent of removal for n-alkanes smaller than n-C_{35}. $\delta^{13}C$ data provide compelling evidence of passive sequestration of crude oil into soil organic matter, but it does not appear that composting stimulated a significant amount of incremental binding. Composting restored soil productivity to pre-impact levels at residual [TPH] of 10,000mg/kg, well above Alberta's 1000mg/kg regulatory guideline. Simulated leachate concentrations satisfied surface runoff objectives throughout the trial, eliminating the need for a synthetic liner. Composting transformed crude oil-impacted soil into CCME unrestricted use compost justifying full-scale (5000m^3) implementation in 1998.

INTRODUCTION

Impacted soil at inactive exploration and production (E&P) properties in Alberta is commonly clay-textured and saline/sodic due to the incidental release of formation brine. Crude oil may comprise several percent of the soil matrix, and may be enriched in high molecular weight n-alkanes via weathering or co-disposal of pigging wax or tank bottoms.

To achieve regulatory closure for inactive E&P properties in Alberta, chemical remediation objectives must be attained and the agronomic capability of the soil restored. The ability of bioremediation to achieve these goals for soil impacted by crude oil and formation brine is complicated by regulatory and technical constraints.

The most difficult regulated guideline to achieve at E&P sites in Alberta is the 1000mg/kg "mineral oil and grease" value (AEP, 1994). This arbitrary limit is inconsistent with recently published risk-based values for crude oil-sourced TPH, which exceed 15,000mg/kg (McMillen et al., 1998). Recognizing that bioremediation generally removes less than 70% of crude oil-sourced TPH (Huesemann, 1995; Visser, 1998; McMillen et al., 1995), the 1000mg/kg guideline in Alberta effectively restricts the utility of bioremediation to soils with starting concentrations below 5000mg/kg, an uncommon occurrence.

In terms of technical constraints, the clay matrix renders desalinization pretreatment technically impractical, and limits mass transfer during bioremediation. Residual salinity inhibits revegetation and, at the extreme, reduces the biotransformation rate (McMillen et al., 1995).

OBJECTIVE

Based upon the generalized inability of conventional land treatment to achieve 1000mg/kg TPH and performance-based revegetation objectives for crude oil-affected soil and, in light of some encouraging groundwork (i.e. McMillen et al., 1996), concurrent composting and land treatment studies were initiated in 1997 to determine whether composting:

1. Enhanced the extent of TPH removal relative to land treatment (with inference as to a possible causative mechanism).
2. Restored soil productivity to pre-impact levels.
3. Produced compost that fulfilled the Canadian Council of Ministers of the Environment (CCME, 1996) specification standards for Category A (unrestricted use) product.

MATERIALS AND METHODS

As part of an aggressive mitigation program, soil from an inactive flarepit in a Paleozoic-sourced oilfield in Alberta was collected for the project. The nature and degree of contamination encountered is typical of impacted E&P pits ([TPH]=25,300±700mg/kg of weathered, paraffinic crude oil; EC = 9dS/m; texture: 18% sand, 30% silt, 52% clay).

The composting trial was conducted at the Olds College Composting Technology Centre between April 1997 and January 1998. Bin (mesocosm) results identified the organic amendment type (thermomechanical pulp waste [pulp fibre and biosolids]) and application ratio (2 soil:3 pulp waste, starting v/v) to compost in a full-scale dynamic windrow (dimensions: 1.2m high x 3.0m wide x 12.25m; starting volume: 25m^3) for a nine-month period. The physical, chemical, and microbiological (bacteria, actinomycetes, fungi) properties of the windrow were monitored routinely to ensure optimal conditions for bioremediation prevailed. Conditions outside predefined control limits prompted amending and/ or turning the windrow as necessary using a self-propelled windrow turner. During the curing phase, maturity and compost classification were assessed using the CCME (1996) protocol. Vegetative capability of the composted soil was assessed relative to the uncomposted soil: pulp waste mixture (positive control) and control soil from the same horizon (negative control) at 6 and 9 months using barley (*Hordeum sp.)* creeping red fescue (*Festuca rubra*), and white dutch clover (*Trifolia repens*). Treatments were performed in triplicate and arranged in the greenhouse in a random block design.

The companion land treatment simulation was performed at the University of Calgary (Visser, 1998). Aliquots of sieved, contaminated soil were incubated under optimal moisture, temperature, and nutrient conditions until a biotreatability endpoint was achieved, as determined by CO_2 evolution (97 days in this case).

TPH concentration was determined using a variation of U.S.EPA 8020 involving dichloromethane (DCM)-soxhlet extraction, silica cleanup (to remove polar plant products) and GC-FID quantitation against a diesel standard. Non-extractable $\delta^{13}C$ was measured via DCM-soxhlet extraction, acidification, and isotope ratio mass spectrometry.

RESULTS AND DISCUSSION

Composting Performance. The log phase of microbial growth was coupled with a rapid generation of exothermic heat and dramatic depression in interstitial oxygen content. Commencing on day 13 the windrow maintained thermophilic temperatures for a 1 week period ($T_{max}=56°C$). In contrast with the protracted thermophilic phase encountered in solid waste composting, the soil: pulp waste windrow plateaued within the mesophilic range for 85 days thereafter ($T_{mean}=30°C$). Over the 105 day thermophilic/mesophilic period, the core temperature exceeded the ambient air temperature by an average 22°C, critical to extending the treatment season in cold climates. Unpublished research on the composting of crude oil impacted soil (McMillen, 1997) suggests that relative to the thermophilic regime, mesophilic temperatures favor greater microbial diversity and realize greater TPH removal.

TPH Removal. Time series [TPH] data is depicted in Figure 1. Simple statistical comparison of the biotreatability endpoints is confounded by the mass dilution induced by the pulp waste in the composting trial. The effect of this dilution is apparent in the instantaneous TPH reduction from 25300mg/kg (source soil) to 18500mg/kg at initiation of the composting trial. Composting the pulp waste by itself enabled estimation of the dilution factor both by measurement and calculation. These independent methods provided a consistent dilution factor of 1.37 over the course of the trial. To meaningfully compare the endpoints attained by composting and land treatment the composting [TPH] data must be adjusted by this factor. The mass dilution-corrected composting [TPH] distribution at 3 months is 15200±1400mg/kg. Compared with the land treatment [TPH] endpoint distribution of 16650±2200, the difference is neither intuitively nor statistically significant. Hence composting was not shown to enhance the extent of TPH removal relative to land treatment in this project. The apparent [TPH] and hopane-normalized [TPH] reductions realized were approximately 40% and 60%, respectively.

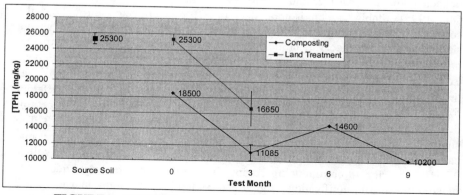

FIGURE 1. [TPH] Time series data (mean ± standard deviation)

n-alkane Depletion. Hopane-normalized [n-alkane] data qualitatively indicate that composting provided superior removal up to n-C$_{35}$ (Figure 2). For larger homologs, no appreciable difference in the extent of constituent removal between composting and land treatment is apparent. Of particular note is the high degree of removal (>55%) realized in both treatments for n-alkanes larger than n-C$_{45}$. Studies involving pure bacterial cultures (e.g. Sugiura et al., 1997; Heath et al., 1997) suggest a molecular-size biotreatability "ceiling" between n-C$_{35}$ and n-C$_{45}$. The enhanced depletion observed in the compost and land treatment systems for n-alkanes larger than n-C$_{45}$ could be attributable to the diverse microbial consortia characterized. The analytical precision of the [n-alkane] data in this range was confirmed. In both treatments, removal exceeded 40% for all constituents.

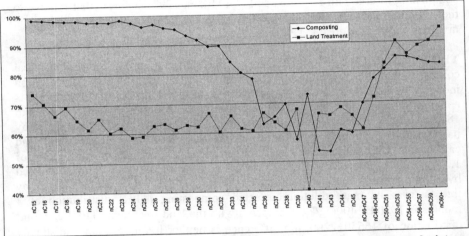

FIGURE 2. % Depletion of hopane-normalized [n-alkane] at biotreatability endpoints

Oxidative Coupling. Several studies have confirmed the importance of biologically- mediated oxidative coupling as a sequestration mechanism for PAH-impacted soil (e.g. Kastner and Mahro, 1996; May et al., 1997; McFarland and Qiu, 1995). Stable carbon isotope data ($\delta^{13}C$) was collected throughout the composting trial to determine if n-alkane dominant substrate (crude oil) induced a similar process. In theory if crude oil constituents are incorporated into native organic matter by a stimulated humification process, the $\delta^{13}C$ of the non-extractable organic carbon in the composting soil should trend towards the $\delta^{13}C$ of the source oil over time. Well-differentiated $\delta^{13}C$ signatures for the different forms of carbon present in the composting system are prerequisites to successful demonstration of this theory. Time series $\delta^{13}C$ data are illustrated in Figure 3. The significant difference between the $\delta^{13}C$ signatures of the non-extractable organic carbon in the *uncontaminated* soil (-22.8) and that of the *contaminated* soil (-26.5±0.26) **prior to commencing** composting provides compelling evidence of passive (intrinsic) humification. Initial and 3-month composting data for Bin 2 (*no* crude oil) and the windrow compost (containing oil) suggest the reduction in $\delta^{13}C$ measured for the windrow can largely be accounted for by mineralization of pulp waste constituents and the resulting increase in the relative amount of mineral soil. Hence oxidative coupling was not shown to have been a significant contributor to the [TPH] reduction realized.

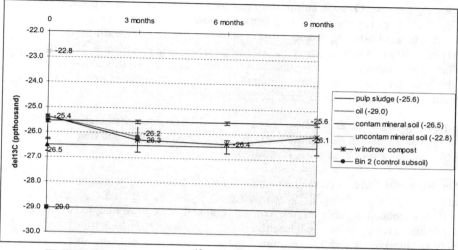

FIGURE 3. Composting δ^{13}C data (mean ± standard deviation)

Vegetative Capability. The mature compost demonstrated an ability to support native vegetation equivalent to that of uncontaminated soil from the same horizon (- control), even though the compost contained ~1% residual [TPH] (Table 1). One exception to the foregoing generalization is the productivity result for fescue, a hardy cultivar with moderate to high salinity tolerance. Nine-month fescue productivity data is inconsistent with 6-month results, in which no significant differences between treatments were observed.

TABLE 1: Vegetation metrics after 9 months (mean±st. deviation)

Treatment	Cumulative Emergence (%)			Final Survival (%)		
	Barley	**Clover**	**Fescue**	**Barley**	**Clover**	**Fescue**
- control	100±0a*	53±0a	80±0a	98±4a	92±7a	89±10a
+ control	100±0a	42±10a	80±12a	100±0a	94±10a	97±6a
Compost	100±0a	47±7a	76±4a	96±4a	83±29a	80±12a

Treatment	Plant Resilience (g dry biomass /plant)			Productivity (g dry biomass/pot)		
	Barley	**Clover**	**Fescue**	**Barley**	**Clover**	**Fescue**
- control	*0.24±0.01b*	*0.017±0.003b*	0.087±0.016a	*3.50±0.10b*	*0.127±0.031b*	*0.93±0.21b*
+ control	*0.20±0.02a*	*0.008±0.004a*	0.083±0.028a	*2.97±0.25a*	*0.053±0.042a*	*0.93±0.21b*
Compost	*0.25±0.02b*	*0.026±0.004c*	0.064±0.013a	*3.63±0.15b*	*0.153±0.065b*	*0.57±0.06a*

*Means followed by different letters within a column are significantly different at P< 0.05 by Fisher's LSD procedure (N = 3).
**Italics* denote measures and species for which statistically-significant differences between treatments were observed

The productivity and resilience of barley and clover grown in the compost were significantly higher that those sustained by the uncomposted soil: pulp waste mixture (+ control), attesting to the vegetative benefit conferred by the composting process. Overall, composting restored soil productivity to pre-impact levels, a necessary requirement for reclamation closure. Of particular interest, this performance-based requirement was achieved at residual crude oil-derived [TPH] of 10200mg/kg, well above the 1000mg/kg regulatory guideline. Note the productivity of the post land treatment soil was not assessed. Hence no conclusions on the relative ability of composting and land treatment to restore agronomic capability can be inferred from this work.

Compost Classification. Physical, chemical, and biological characterization of the compost at maturity confirmed that the final product fully satisfied CCME (1996) guidelines for Category A (unrestricted use) compost.

Other Relevant Parameters. Priority PAH concentrations were reduced from a sum of 1.0mg/kg to 0.4mg/kg over the 9 month composting trial. At each 3-month milestone a sample of the composting soil was saturated to simulate precipitation events. Simulated leachate concentrations satisfied surface runoff objectives throughout the trial, eliminating the need for a synthetic liner at full scale.

CONCLUSIONS
- Collectively, the [TPH] and [n-alkane] data for this project suggest that composting and land treatment were equally effective in biotically removing crude oil constituents. The apparent [TPH] reduction achieved by both treatments was approximately 40%.
- Composting enhanced the extent of removal for n-alkanes smaller than $n\text{-}C_{35}$. For larger homologs, no appreciable difference in the extent of constituent removal between composting and land treatment was apparent. Of particular note is the high degree removal (>55%) realized in both treatments for n-alkanes larger than $n\text{-}C_{45}$.
- $\delta^{13}C$ data provide compelling evidence of passive (intrinsic) humification of crude oil, but it does not appear that composting stimulated a significant amount of incremental binding.
- Composting restored soil productivity to pre-impact levels at residual [TPH] $\cong$10,000mg/kg. Coupled with McMillen et al. (1998) guidance on human health-protective [TPH], the collective data set suggests the 1000mg/kg Alberta guideline may be unnecessarily conservative for crude oil-sourced TPH.
- Composting transformed crude oil-impacted soil into CCME Category A (unrestricted use) compost.
- Composting offers some unique benefits: in cold climates, mesophilic temperatures extend the duration of favourable bioremediation conditions,

potentially permitting year-round treatment; it also restores soil productivity, a prerequisite to regulatory closure in Alberta.
- Progressing to full scale was justified by the demonstrated ability to produce Category A compost. As a result, full-scale composting was implemented in the summer of 1998, comprising $5000m^3$ of material configured in windrows totaling 1.7km in length. Lifecycle costs of the project are estimated at $80 per m3 of remediated soil.

ACKNOWLEDGEMENTS

The authors would like to thank Sara McMillen and Michael Moir of Chevron for providing invaluable assistance in designing the trial and pragmatic scale-up suggestions. We are also grateful to Peter Jenden and Bob Carlson of Chevron for their expert geochemical advice.

REFERENCES

Alberta Environmental Protection (AEP). 1994. *Alberta Tier I Criteria for Contaminated Soil Assessment and Remediation.* Edmonton, Alberta.

Canadian Council of Ministers of the Environment (CCME). 1996. *Guidelines for Compost Quality.* Ottawa, Ontario. CCME 106E.

Heath, D.J., C.A. Lewis, S.J. Rowland. 1997. "The use of high temperature gas chromatography to study the biodegradation of high molecular weight hydrocarbons". *Org. Geochem.* 26(11-12): 769-785.

Huesemann, M.H. 1995. "A Predictive Model for Estimating the Extent of Petroleum Hydrocarbon Biodegradation in Contaminated Soils". *Environ. Sci. Technol.* 29(1): 7-18.

Kastner M., and B. Mahro.1996. "Microbial Degradation of Polycyclic Aromatic Hydrocarbons in Soils Affected by the Organic Matrix of Compost." *Appl. Microbiol. Biotech.* 44: 668-675.

May, R., P. Schroder and H. Sandermann. 1997. "Ex-Situ Process for Treating PAH-Contaminated Soil with *Phanerochaete chyrsosporium*". *Environ. Sce. Technol.* 31(9):2626-2633.

McFarland, M.J. and X.J. Qiu. 1995. "Removal of benzo[a]pyrene in soil composting systems amended with the white rot fungus *Phanerochaete chrysosporium*". *Journal of Hazardous Materials.* 42: 61-70.

McMillen, S.J., A.G. Requejo, G.N. Young, P.S. Davis, P.D. Cook, J.M. Kerr and N.R. Gray. 1995. " Bioremediation Potential of Crude Oil Spilled on Soil". In R.E. Hinchee, F.J. Brockman and C.M Vogel (Eds.), *Microbial Processes for Bioremediation*, pp. 91-99. Battelle Press, Ohio.

McMillen, S.J., J.M. Kerr, P.S. Davis, J.M Bruney, M.E. Moir, P. Nicholson, C.V. Qualizza, R. Moreau and D. Herauf. 1996. *Composting in Cold Climates: Results from Two Field Trials.* Presented at the SPE International Conference on Health, Safety and the Environment. New Orleans, LA. 9-12 June 1996.

McMillen, S.J. 1997. Personal communication regarding unpublished research on composting crude oil impacted soil.

McMillen, S.J., G. Deeley and J. Kerr. 1998. *A Method for Establishing Total Petroleum Hydrocarbons (TPH) Risk-Based Screening Levels at Upstream Sites.* Presented at the 5th International Petroleum Environmental Conference, Albuquerque, NM.

Sugiura, K., M. Ishihara, T. Shimauchi and S. Harayama. 1997. "Physicochemical Properties and Biodegradability of Crude Oil". *Environ. Sci. Technol. 31*: 45-51.

Visser, S. 1998. *Flarepit Remediation: Use of Ecotoxicity, Biodegradation and Waste Chemistry for Pit Classification and Prediction of Treatability – Progress Report.* Prepared for Canadian Association of Petroleum Producers and Environment Canada.

WASTE COMPOUND DEGRADATION IN GASES AND HEAVY METAL CONTAINING WATER

Hans-Peter Schmauder (fzmb, Bad Langensalza, Germany), Jan Paca (Inst. Chem.Technol., Prague, Czech. Republic), Jerzy Dlugonski (Univ. Lodz, Poland), Sava Mutafov (Inst. Microbiol., Sofia, Bulgaria), Francisco Girio, José Menaia, M. Rosário Freixo (all: INETI, Lisbon, Portugal) and A. Mark Gerrard (Univ. Teesside, Middlesbrough, England)

ABSTRACT: Two joint research projects of the European Community deal with the biotechnological control of environmental pollutants in soil, water and gases mainly for an application in the countries of Central and Eastern Europe. The main studied pollutants were among others styrene, benzene, toluene, ethylebenzene, and xylenes (BTEX); trichloroethene (TCE); polycylcic aromatic hydrocarbons (PAHs), and phenolic compounds. Some species of bacteria (Pseudomonas, Comamonas, Acinetobacter, Xanthomonas) and fungal strains (Verticillium marquandii) were isolated from contaminated areas in Bulgaria, Poland, the Czech Republic, and Portugal. On the examples heavy metal accumulation and enhancing naphthalene bioavailability the results were described in detail. Waste compound degradation also is effective using immobilized microbial strains. Modelling experiments also were running during these two joint research projects.

INTRODUCTION

Humans have emitted pollution into the environment for centuries. In earlier times, the pollutants were of mostly natural origin, but since industrialization they have gradually changed to substances that pose increasing difficulties for biological degradation. Particularly in Central and Eastern Europe, such pollution will become more and more problematic. For this reason, the European Community has commissioned two joint research projects for the biotechnological control of such pollutants in gases and water in areas with especially high concentrations of heavy metals.

Our work includes: Analyses of the actual levels of pollution, researching bench scale treatment possibilities and developing new, larger scale methods; Studying the purification of gases (polluted with styrene and related compounds, trichloroethylene [TCE] and benzene, toluene, ethylbenzene and xylenes [BTEX]), by using new isolated strains (genus Comamonas or Pseudomonas) or mixed populations in free state or immobilized; Isolating and selecting microorganisms which tolerate higher concentrations of heavy metals and have an ability to degrade waste compounds; and the optimal economic design and operation of biofilters and/or bioreaction systems for degradation of toluene and xylene as model compounds.

MATERIALS AND METHODS

Microorganisms and Their Isolation.   Verticillium marquandii IM 6003 was used that had been isolated in the Department of Industrial Microbiology, Institute of

Microbiology and Immunology, University of Lodz (Poland) from samples of solid flotation tailings collected in Silesia region (Poland).

The selective isolation of toluene- and xylene-assimilating bacteria was performed in continuous flow conditions. For this purpose a modified cultivation technique resembling the C-limited pH-auxostat (Büttner et al., 1986) or bistat (Minkevich et al., 1987) was used (Figure 1). The modification made concerned the addition of the C-substrate (toluene or xylene) which was fed into the cultivation space via the air flow. All other components were fed via the pH-controlled liquid flow. The medium was added as two separate solutions by the two-channel peristaltic pump [15]. The composition of the solution fed via channel 1 was (mg/L): KH$_2$PO$_4$ 1,000, MgSO$_4$.7H$_2$O 500, NH$_4$Cl 800, FeSO$_4$.7H$_2$O 16, CaCl$_2$.6H$_2$O 5, CuSO$_4$.5H$_2$O 0.15, CoCl$_2$.6H$_2$O 0.18. Via channel 2 2.5% NH$_4$OH solution was fed. The flow ratio between channel 1 and channel 2 was 8.4:1. The source of isolation was river water from Bulgaria. The isolation was carried out at 32°C, pH 6.8, aeration 60 L/h, and stirring 1,100 min^{-1}.

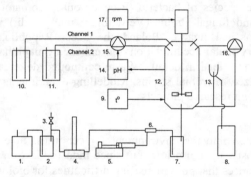

Fig. 1. The equipment scheme. 1. Air pump; 2. Damper; 3. Needle valve; 4. Air-flowmeter; 5. Syringe pump; 6. Evaporator; 7. Damper; 8. Container for effluent culture liquid; 9. Temperature control module; 10. Ammonium solution container; 11. Salt solution container; 12. Fermenter; 13. Level control; 14.-15. pH-control with two-channel peristalticpump; 16. Peristaltic pump; 17. Stirrer control module.

The isolates were determined taxonomically by BIOLOG system. Toluene assimilating microorganisms were Acinetobacter calcoaceticus (baumannii) Gen. 2, A. radioresistens genospecies 12 and Xanthomonas maltophilia. Xylene assimilating microorganisms were Xanthomonas oryzae PV E, Acinetobacter genospecies 9, Pseudomonas diminuta, P. putida type Al and P. fluorescens type G.

Other used microorganisms were isolated from different sources of soils and water in Germany, Portugal, and the Czech Republic. Both kinds were applied during the experiments, mixed cultures as well as pure strains of the genera Pseudomonas, Bacillus, Comamonas etc.

Culture Conditions, Heavy Metals Uptake and Analytical Procedures.

Conidia were washed off a 14 days old culture on Sabouroud slants and 1 ml suspension (containing $1x\ 10^8$ conidia) was transferred into 19 ml of Sabouraud liquid medium in 100 ml flask. After 24 h of incubation on rotatory shaker (28°C, 160 rpm) the preculture was introduced into one-liter flask with fresh medium (in the ratio 1:9) supplemented with heavy metals salts (0.5 g/L) and incubated under the same conditions for 7 days. The samples of mycelium were collected on Sartorious filters (0.8 µm) and hplc will be made.

RESULTS AND DISCUSSION

Heavy Metal Uptake.   The results shown in Fig. 2 indicate that the presence on zinc and lead compounds in Sabouraud medium has no inhibitory effect on growth of Verticillium marquandii IM 6003. Contrary to the carbonates the heavy metals added as acetates (Fig. 2) seem to stimulate the fungal growth most probably as a consequence of acetate utilization. The investigated strain was isolated from samples of industrial dumps collected near non-ferrous metal work in Silesia region (Poland), an area contaminated by heavy metals. The bacterial strains isolated form the samples collected in this area, too, also were able to grow at high concentration of zinc and lead (Fijalkowska et al., 1998). It is possible that the observed resistance of the isolated strains is combined with adaptation of microorganisms to the polluted environment.

The analyses of zinc and lead uptake by growing cultures of V. marquandii IM 6003 (Fig. 3) indicate that the studied strain is able to accumulate zinc and lead added to the cultures in the form of easily (acetates) and poorly (carbonates) soluble salts. The best heavy metals accumulation ability was observed in the stationary phase of growth and is comparable with uptake ability of fungal strains used for heavy metal removal from aqueous solutions (Volesky and Holan 1995). Factors controlling heavy metals biosorption by V. marquandii IM 6003 mycelium are under study.

Enhanced Naphthalene-Bioavailability in a Liquid-Liquid Biphasic System.

A PAH- degrading bacterial co-culture (EXPO98) was isolated from a soil with a heavy and old crude-contamination. EXPO98 was used in batch studies to investigate comparatively the effects of particle size, the presence of an apolar phase, and the addition of surfactants on the rates of naphthalene degradation and coupled microbial growth. Growth on naphthalene, added as a single particle or dispersed by sonication, in solution in dimethylpolysiloxane, and in the presence of Tween 80 or Triton X-l00, was exponential as far as naphthalene was detectable in the culture medium.

Linear growth occurred thereafter. While observed exponential growth rates were similar for all tested systems, the length of logarithmic growth was longer on dispersed naphthalene and was significantly extended when dimethylpolysiloxane was present as an organic phase. In addition, subsequent zero-order growth was

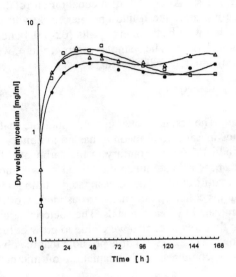

FIGURE 2. The growth curve of *Verticillium marquandii* IM 6003 in presence of zinc acetate and lead acetate. • - without heavy metals (control), □ - with zinc acetate, △ - with lead acetate

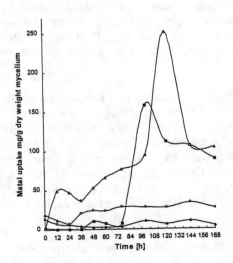

FIGURE 3. Zinc and lead uptake by *Verticillium marquandii* IM 6003 in growth medium with supplements of heavy metal salts. Supplements used: ■ - zinc carbonate, □ - zinc acetate, ▲ - lead carbonate, △ - lead acetate.

significantly faster in the biphasic system. Although EXPO98 was able to grow on Tween 80 as the sole substrate, surfactant addition to cultures on solid naphthalene did not produce any detectable effect on. After depletion of the starting equilibrium concentration of aqueous naphthalene PAH bioavailability was controlled by its rate of dissolution.

Moreover, it was likely that the transfer of naphthalene to water was faster from the organic than from the solid phase, provided that an adequate concentration-gradient remained between the organic and aqueous phases. Microscopic observations and investigation of cell distribution showed that cells were preferentially suspended in the aqueous phase, so weak cell-hydrophobicity may not exclude the useful utilization of microorganisms in liquid-liquid biphasic systems. The presence of an organic phase facilitated the PAH bioavailability. As many hydrophobic compounds easily and rapidly separate from the bulk culture medium, their use may lead to biotechnological developments on continuous biphasic-systems with recycling of the liquid organic-phase (which may help to overcome present rate limitations in the utilization of microorganisms to remove PAH contamination).

Degradation of Benzene, Toluene, Phenols and the Application of Detergents and Immobilized Cells. A further main focus was given to aspects of an increase of solubility of the partially poor water soluble substrates, e.g. BTEX, PAH, styrene, and the application of immobilized microbial systems in the processes. The used detergents were mainly types of AERES T22, T27 and related mixtures (see also Angelova and Schmauder [1999] and Schmauder et al. [1997]. The main immobilization processes were done using expanded slate, other inorganic materials (e.g. *Terraperl* S and *Öl-Ex hart)* or some organic polymers (polystyrene) or peat materials (for details see also: Angelova and Schmauder [1999], Paca et al. [1999]). The results and first modelling experiments demonstrate that the application of biodegradable detergents and of immobilized microbial systems will be helpful for a solution of open questions. Further experiments on this topic are in progress.

REFERENCES

Angelova, B. and Schmauder, H.-P. (1999). "Lipophilic Compounds in Biotechnology - Interactions with cells and Technological Problems." *Journal of Biotechnology* 67: 13-32.

Biittner, R., B. Uebel and I. Genz. 1986. "The Substrate Limited phAuxostat: a New Method for Continuous Culture." *Acta Biotechnol.,* 6 (2): 129-132.

Fijalkowska S., Lisowska K. and Dlugonski J. 1998. "Bacterial Elimination of Polycyclic Hydrocarbons and Heavy Metals." *Journal of Basic Microbiology* 38: 361-369.

Minkevich, I. G., A. Y. Krynitskaya, and V. K. Eroshin. 1987. (in Russian). "The Bistat (Chemostat Plus pH-Auxostat) Method of Cultivation Studied by Growth of *Candida valida* on Ethanol". *Mikrobiologiya,* 56 (2): 254-258.

Paca, J., Weigner, P., Koutsky, B., Metris, A. V., Gerrard, A. M., Frankenfeld, K. and Schmauder, H.-P. 1999. "Aerobic Waste Compound Degradation in Packed Bed Reactors." In *Proceedings of the Fifth Int. Symposium ,,In Situ and On-Site Bioremediation,* Battelle, in press.

Schmauder, H.-P., Frankenfeld, K., Ludwig, M. and Luthardt, W. 1997. "Purification of Contaminated Oil-binding Materials for Repeated Use." In B. C. Alleman and A. Leeson *In Situ and on-Site Bioremediation* Vol. 5, pp. 259-263, Battelle Press, Columbus, Richland

Volesky B. and Holan, Z. R. 1995. "Biosorption of Heavy Metals". *Biotechnol. Prog.,* 11: 235-250.

Acknowledgement: The authors are grateful for grants from the EC for this work (ERBIC 15CT96/0710, ERBIC 15CT96/0716) and from national governments.

BIOLOGICAL TREATMENT OF WASTEWATERS FROM PESTICIDE AND STARCH MANUFACTURE

*J. (Hans) van Leeuwen**, Richard U. Edgehill** and Bo Jin
University of New England, NSW, Australia

*Presently Visiting Professor at Civil and Environmental Engineering,
San Diego State University CA 92182-1324
**Presently at San Joaquin Valley Unified Air Pollution Control District,
Bakersfield CA 93301-2370

ABSTRACT: Treatment processes were developed for two different industrial wastewaters, both with a high COD. Both processes involved aerobic treatment and removed substantial quantities of pollutants. The similarity ends there, as the microbial populations and the results differed in almost every possible way.

The *pesticide* wastewater had a high COD and contained a variety of pesticides, mainly endosulphan 1 and 2, diazinon, malathion, atrazine, simazine, chlorpyrifos and others in a total concentration of c. 1.6 g/L. Normal disposal methodology involves encapsulation and costs several thousands of dollars per cubic meter. The bench-scale research described in this paper involved ozonation, biological granulated activated carbon treatment, biofilter treatment only and chemical coagulation with microfiltration. A process without activated carbon was studied in parallel. Both the biofilter and biological activated carbon with coagulation and separation with microfiltation were able to remove about 99% of most pesticides, with the bulk of the removal occurring during the biological treatment stage. A notable exception was simazine, which left persistent residues. While ozonation improved the COD removal, it was not deemed an essential treatment component. The effluent would have been suitable for sewer discharge with costs two orders of magnitude less than encapsulation.

The *starch* wastewater had a COD of 12300 mg/L (of which 7900 mg/L soluble), a BOD of c. 9600 mg/L, a TOC of 4550 mg/L and presented a disposal problem mainly as high sewer discharge fees to the manufacturer. A single, nonaseptic fungal treatment process was developed, which not only removed 95% of the BOD and 78 to 85% of the total organic carbon (TOC), but produced a biomass product with a protein content of 38 to 48%, suitable for animal feeds. Different strains of *Aspergillus oryzae* and *Rhizopus arrhizus* were found to be suitable, but the efficiency of conversion of waste to fungi was substantially influenced by the species and strain of organism selected. The yield achievable was 1.3 to 1.44 g biomass/g TOC. Addition of plant nutrients was not essential, but some increase in biomass production could be achieved with phosphate, magnesium and calcium supplementation. An air lift reactor configuration was developed with air spargers and external recirculation resulting in velocity gradients which favored the growth of pelletized colonies. These are easily harvested and dewatered by screening. Further dewatering could be done with a simple process such as belt pressing.

The product has a potential value of $200-300 per ton, suggesting a profitable venture.

INTRODUCTION

Australia has a very unique, but extremely fragile ecology. Wastes and wastewaters have to be disposed of in very responsible methods. There is concern in Australia about the increasing use of pesticides for crop protection (Schofield *et al.*, 1998). Pesticide manufacturing wastewater (PMW) is produced during rinsing and wash-down operations at pesticide production facilities and is contaminated with pesticides and other chemicals. Because of the highly heterogeneous chemistry of PMW, treatment becomes particularly challenging and may involve the use of more than one technology. In this work, the potential of ozonation, filtration, coagulation, activated carbon adsorption, and biofilter treatment for removal of pesticides was examined on laboratory scale.

Australia converts wheat flour into starch in most of the major cities in order to provide a raw material for the manufacture of a variety of adhesives as well as applications within the food industry. The process, which is essentially a wet one, produces substantial quantities of wastewater with high concentrations of organic material. Whilst this material is relatively innocuous, it needs to be treated biologically to lower the BOD before disposal. One company was paying millions in sewage disposal charges per annum because of this high BOD and has spend more millions in installing an on-site anaerobic treatment process at one of their plants. The research described in this paper was aimed at substantially lowering this BOD, while also producing a useful proteinaceous biomass. The investigation was done on bench scale to develop parameters for the design of a pilot plant.

MATERIALS AND METHODS – PESTICIDE WASTEWATER TREATMENT

Wastewater samples were obtained from a pesticide manufacturing facility. Undiluted PMW was supplemented with 13 g/L $NaNO_3$ and 0.5 g/L KH_2PO_4 which lowered its pH to c. 7. PMW was ozonated using an Oztech ozone generator. The dosage was about 20g/L.

Biofilm treatment of PMW was evaluated in columns of dimensions 400 mm high and 100 mm in diameter containing gravel (10 mm basalt). The gravel columns were inoculated with microorganisms obtained from a sewage treatment plant acclimated to diluted wastewaters, gradually increasing its concentration in recycling batches of wastewater gradually increased up to full strength. Repetitive batches of 500 mL of PMW were recirculated for a week each through this column at a rate of 100mL/h. After four months, 2/3 of the gravel in one of the columns was replaced with approximately 640 g activated carbon pellets (Model A-38-1 Catic, Beijing, China).

Ferrous sulfate was used to coagulate the laboratory-biotreated PMW. The wastewater was dosed with 2 g/L Fe(II), added as solid $FeSO_4.7H_2O$, stirred manually for mixing and coagulation, course filtered and then adjusted to pH 9-10 with NaOH. The precipitate was removed by $0.45\mu m$ membrane filtration.

The wastewater was analyzed for pesticides, total organic carbon (TOC), total solids and total suspended solids, biochemical oxygen demand (BOD), chemical oxygen demand (COD), chloride, alkalinity, and electrical conductivity. TOC was analyzed using a Tekmar-Dohrmann DC-190 organic carbon analyzer. Quantitative analyses for organochlorine (OC), organophosphorus (OP), phenoxy herbicides, and triazine pesticides were done using GC-MS by Australian Laboratory Services P/L (Brisbane, Queensland, Australia). Carbamate pesticides were analyzed by the Queensland Government Chemical Laboratory. Biofilms were examined with a JEOL 5800LV scanning electron microscope after osmium vapor fixation for several hours and sputtering the specimens with gold. Glass cover slips (Menzel-Glaser, Germany) were added to the cultures and removed at various times for observation.

MATERIALS AND METHODS – STARCH WASTEWATER TREATMENT

Thirty-two strains, consisting of 13 fungal species from *Fusarium, Trichoderma, Aspergillus, Chaetomium, Geotrichum, Rhizopus* and *Paecilomyces,* and 4 yeast species from *Candida, Saccharomyces (or Endomyces), Endomycopsis* and *Aureobasidium,* were tested in this project. All cultures were obtained from the Culture Collections of the Department of Primary Industries, Queensland and the CSIRO Division of Food Science & Technology, and the Biological and Chemical Research Institute in Australia in 100 mL batches of starch wastewater cultivated in 250 mL Erlenmeyer flasks (Jin *et al.*, 1999a). Suitable candidates were selected on the basis of efficiency of TOC conversion, biomass production and rate of growth.

Different bench scale air lift reactors were operated with 2 strains of *Rhizopus arrhizus* and one of *Aspergillus oryzae* on semi-batch and continuous basis using starch wastewater as feed (Jin *et al.*, 1999b). Tests were conducted on the effect of the addition of plant nutrients nitrogen and phosphate, optimal temperature and pH for growth, the effect of the trace elements calcium, magnesium, iron and zinc. The effect of different reactor configurations on the morphology of the biomass and the effect this had on the separation of effluent from the biomass were also evaluated.

RESULTS AND DISCUSSION – PESTICIDE WASTEWATER

PMW was dark green in color, had a putrid odor, and contained solid and dissolved pesticides, solvents, surfactants, and other chemicals.

Ozonation removed both color and odor from the wastewater. Ozone added at 0.47g ozone/g COD removed c. 20% of the COD.

Biological treatment changed the color to brown and the odor to much less putrid. Biological treatment on its own could remove 88 to 95% of the COD. Pre-ozonation improved subsequent removal of COD by biological treatment to achieve a 95% removal. In all cases subsequent filtration was required to separate colloidal material from the effluent and values reported include the action of filtration.

Coagulation of the wastewater removed COD, TOC, color and turbidity. Ferrous sulfate coagulation followed by membrane filtration reduced the COD of the PMW to 3560mg/L and the TOC to 811 mg/L, ie a 95% reduction in both COD and TOC. Large quantities of sludge were generated, however, which would create a further disposal problem, albeit on a smaller scale.

Activated carbon adsorption removed approximately 80% of the COD from the three effluents: coagulated PMW, coagulated biotreated PMW, and ozonated PMW at an equivalent "dose" of 30 g/L. The activated carbon was capable of removing more than 1.5 g COD/g carbon from ozonated PMW on a cumulative basis. Much of this could be ascribed to biological action. Microscopic investigation also confirmed a much more dense population of bacteria in the column treating ozonated PMW. The flocculated and motile rod-shaped bacteria occurred in thicker layers and contained more extracellular polymer. Enhanced removal of organic material after pre-ozonation of wastewater was observed earlier by Van Leeuwen *et al.* (1983) and Heinzle *et al.* (1992). The COD removal efficiency remained fairly constant throughout the study indicating that saturation of the carbon had not occurred and that biodegradation was also an important removal mechanism (Suidan *et al.*, 1980; van Leeuwen *et al.*, 1983).

Microfiltration using 0.45 µm cellulose nitrate membranes removed 86% and 90% COD and TOC, respectively, from PMW and 23% of the COD from biologically treated PMW. However, only small volumes (< 5 mL) of PMW could be filtered because of clogging.

Pesticide removal by biological treatment only and by biological activated carbon on sequential batch basis is shown in Table 1. Both processes removed most of the pesticides. Simazine was the only one that could not be removed effectively. Concentrations approached values that made discharge into a sewer permissible.

Savings in wastewater disposal of more than 95% could potentially be achieved after this level of removal of pesticides and organic material, although more research is still required.

RESULTS AND DISCUSSION – STARCH WASTEWATER TREATMENT

A simple, nonaseptic, low-cost single process was developed for the production of microbial biomass protein (MBP) and wastewater purification from starch processing wastewater (SPW). Three species and six strains of filamentous microfungi *Aspergillus oryzae*, *Rhizopus oligosporus* and *Rhizopus arrhizus* demonstrated high microbial and enzymatic activities on SPW, and a desirable suitability for MBP production using SPW as a sole carbon and energy source. *A.*

oryzae 3863, *R. oligosporus* 2710 and *R. arrhizus* 36017 were representatively selected for the studies.

TABLE 1 Improvement in Wastewater Quality with Biological and Biological Activated Carbon Treatment (Concentrations in mg/L)

Substance/determinant	PMW	Biological Treatment	BAC
Endosulphan 1	1036	8.0	0.116
Endosulphan 2	396	3.5	Nd
Diazinon	17	0.22	0.048
Malathion	74	0.37	
Chlorpyrifos	76	4.76	0.0027
Heptachlor epoxide		0.191	
Atrazine		0.043	
Simazine	nd	9.8	6.21
Dicamba	1.3	0.224	
2,4-D		0.026	
2,4-DB		0.154	
Trichlopyrifos		0.170	
MCPA		0.364	
Carbamate pesticides	<0.1	<0.1	
COD	70,000	8000	
TOC	18,000	2000	
total solids	40,000		
total suspended solids	19,400		
Chloride	2,300		
Alkalinity (CaCO3)	3,600		

A. *oryzae* 3863 exhibited the best results in biomass protein production and TOC reduction at 35°C and initial pH 5.0 of SPW inoculated with 5.0% (v/v) of spore suspension. *R. oligosporus* 2710 and *R. arrhizus* 36017 had optimal culturing conditions at 30°C and initial pH 3.5 and 4.0, respectively with an inoculum size of 7.5% (v/v). The supplementation of ammonium sulphate or urea in the range of 0 to100 mg/L to SPW did not significantly affect the biomass production for the three microfungi, but significantly affected the protein content of *A. oryzae* 3863 and *R. oligosporus* 2710.

The addition of sodium nitrate did not affect either biomass production or protein content. The supplementation of KH_2PO_4 or K_2HPO_4 at 40 mg/L corresponded with a slight increase in biomass production and protein content for the three cultures. KH_2PO_4 was found to be the most suitable phosphorus source. The supplementation of $MgSO_4$ or $CaCl_2$ up to 15 to 20 mg/L led to a 10 to 15% biomass increase in *R. oligosporus* 2710 and *R. arrhizus* 36017 production and to a protein content increase of 10-15% with *A. oryzae* 3863 and *R. oligosporus*

2710. FeSO$_4$ and ZnSO$_4$ added in a range 0 to 15 mg/l did not significantly affect MBP in either quality or quantity, but the addition of ZnSO$_4$ above 15 mg/l would inhibit the microbial growth.

The three microfungi grew exponentially with no pronounced lag phase shorter than 6 h. Exponential growth occurred for c. 5h during the cultivation of *A. oryzae* 3863, *R. oligosporus* 2710 and *R. arrhizus* 36017 at μ= 0.18 h^{-1}, 0.14 h^{-1} and 0.11 h^{-1}, respectively, and the maximum biomass growth was achieved at a biomass production rate of 0.45, 0.35 and 0.23 (g/L/h), respectively, after a cultivation period of 12, 14 and 18 h, respectively. The cultivation of the three microfungi on SPW achieved a biomass yield of 1.30 to 1.44 g of dry biomass/g TOC, a protein yield of 0.55 to 0.67 g of protein/g TOC, and 78 to 85% TOC reduction. *A. oryzae* 3863 and *R. oligosporus* 2710 can nearly completely hydrolyse the starch materials, and *R. arrhizus* 36017 hydrolysed 93% starch over 14 h cultivation. In both shake flask and bioreactor cultures, *A. oryzae* 3863 growth on SPW mainly formed clumpy mycelia and compact pellets, and dispersed fluffy and coalesced mycelia in most cases of growing *R. arrhizus* 36017. (Jin *et al.*, 1999c). Compact pellets, and clumpy and coalesced mycelia are the desirable morphological forms for oxygen transfer and biomass harvesting. These morphological forms could be produced by using precultures as an inoculum and/or controlling the growth pH at 5.0 to 6.0 for *A. oryzae* 3863 and 3.5 to 4.5 for *R. arrhizus* 36017 cultures at high shake rates or high air flow rate.

Both newly designed internal and external air lift reactors (ALRs) are very suitable for the cultivation of the fungal cultures. The diameter ratio of the downcomer and riser D$_d$/D$_r$ =0.71 was determined to be optimal for both ALRs. A gas separator height of 20 to 30 cm was experimentally determined for the operation of laboratory scale ALRs. The newly designed ALRs with a large diameter sparger in the internal ALR and double spargers in the external ALR exhibited a considerably high gas holdup and oxygen transfer coefficient at a given gas flowrate. *A. oryzae* 3863 and *R. oligosporus* 2710 were finally selected as the most suitable strains for MBP production. Using the new ALRs and precultures in the batch cultures of *A. oryzae* 3863 and *R. oligosporus* 2710, the generation time was shortened by 2h compared with the shake flask cultures, and maximum biomass growth was achieved within a period of 10 and 12 h, respectively. An air flowrate of 0.5 to 1.5 v/v/m was required by stages to maintain the dissolved oxygen level above 50% of the saturation during the cultivation. The external ALR with double spargers appeared more suitable for cultivating the selected filamentous microfungi. The selected microfungi could be cultivated successfully in batch, semicontinuous and continuous process, but the semicontinuous mode might be most suited to achieve a high productivity and better quality control. Contamination occurring after a long term run (6-10 batches) of the nonaseptic process might be reduced by running the new external ALR equipped with a cross-flow microscreening unit. The ALRs had a high bioconversion efficiency of starch materials and a short hydraulic retention time requirement, resulting in producing 4.5 to 6.0 g/L of dry biomass. The fungal biomass products contained 38% to 48% protein and may be safe for human and animal consumption.

Nearly complete removal of suspended solids and 95% COD and BOD make the effluent suitable for farm irrigation or substantially reduce sewer disposal costs. All the results can be achieved in running the process without pretreatment of SPW, without nutrient supplementation, and without aseptic operational conditions. This process could be implemented on a large-scale commercial process and it is now being tested on pilot scale by Weston Bioproducts, Melbourne, Australia. Unlike bacterial biomass as used in conventional wastewater treatment processes, this biomass has a value of $200-$300 per ton, which would not only off-set treatment costs, but make this an opportunity to generate a profit.

REFERENCES

Heinzle, E., F. Geiger, M. Fahmy, and O.M. Kut. 1992. Integrated Ozonation-Biotreatment of Pulp Bleaching Effluents Containing Chlorinated Phenolic Compounds. *Biotechnol Prog* 8: 67-77.

Jin, B, J. van Leeuwen, , Q. Yu, and B. Patel. 1999a. Screening and selection of microfungi for microbial biomass protein production and water reclamation from starch processing wastewater. *Journal of Chemical Technology and Biotechnology*, 74:000.

Jin, B, J. van Leeuwen, and H.W. Doelle. 1999b. The influence of geometry on hydrodynamic and mass transfer characteristics in a new external airlift reactor for the cultivation of filamentous fungi. *World Journal of Applied Microbiology and Biotechnology*, in press.

Jin, B, J. van Leeuwen, and B. Patel. 1999c. Mycelial morphology and fungal protein production from starch processing wastewater in submerged cultures of *Aspergillus oryzae. Process Biochemistry*, in press.

Schofield, M.T., V. Edge, and R. Moran. 1998. "Minimizing the Impacts of Pesticides on the Riverine Environment, Using the Cotton Industry as a Model." *Water* (Australia) Jan-Feb issue.

Suidan, M. T., W.H. Cross, and M. Fong. 1980. "Continuous Bioregeneration of Granular Activated Carbon during Anaerobic Degradation of Catechol." *Prog Wat Tech*12:203-214.

van Leeuwen, J., J. Prinsloo, and R.A. van Steenderen. 1983. "The Optimization of Biological Carbon in a Water Reclamation Context." *Ozone Sci & Engng* 5:171-181.

INDUSTRIAL DYE DECOLORIZATION BY LACCASES FROM LIGNINOLYTIC FUNGI

Rafael Vazquez-Duhalt (Instituto de Biotecnología UNAM, México)
Elizabeth Rodríguez (Instituto de Biotecnología UNAM, México)
Michael A. Pickard (University of Alberta, Canada)

ABSTRACT: White-rot fungi were studied for the decolorization of twenty-three industrial dyes. Laccase, manganese peroxidase, lignin peroxidase and aryl alcohol oxidase activities were determined in crude extracts from solid state cultures of sixteen different fungal strains grown on whole oats. All *Pleurotus ostreatus* strains exhibited high laccase and manganese peroxidase activity but highest laccase volumetric activity was found in *Trametes hispida*. Only laccase activity correlated with the decolorization activity of the crude extracts. Two laccase isoenzymes from *Trametes hispida* were purified and their decolorization activity was characterized.

INTRODUCTION

Industrial dyes can be released into the environment from two major sources: as effluents from synthesis plants, and from dye-using industries, such as textile factories. It is estimated that between 10 to 15% of the total dye used in the dyeing process may be found in wastewater (Brown et al., 1981). Several of these dyes are very stable to light, temperature and microbial attack, making them recalcitrant compounds (Pasti-Grigsby et al., 1992).

The white rot fungus *Phanerochaete chrysosporium* is able to decolorize several industrial dyes (Kirby et al., 1995), and polymeric dyes (Glenn and Gold, 1983). *P. chrysosporium* cultures, extracellular fluid and purified lignin peroxidase were able to degrade crystal violet and six other triphenylmethane dyes by sequential N-demethylations (Bumpus et al., 1985). Azo dyes Orange II, Tropaeolin O and Congo Red, and the heterocylic dye Azure B were decolorized by cultures of *P. chrysosporium* (Cripps et al., 1990). However crude lignin peroxidase decolorized all the dyes, except Congo Red, indicating that other enzymes must be involved in the degradation of this azo dye *chrysosporium* (Cripps et al., 1990). The role of purified lignin peroxidase in the decolorization of several azo dyes has been clearly demonstrated (Platt et al., 1985). The different isoenzymes of lignin peroxidase produced by *P. chrysosporium* are able to decolorize several dyes with different chemical structures, including azo, triphenylmethane, heterocyclic, and polymeric dyes (Ollikka et al., 1993). This decolorization was enhanced by the presence of veratryl alcohol. Manganese peroxidase from *P. chrysosporium*, was also able to decolorize several azo dyes *in vitro*, and with both enzymes the decolorization rate was dependent on the chemical structure of the dye (Paszczynski and Crawford, 1991).

Other ligninolytic fungi have shown the capacity for dye decolorization. *Pleurotus ostreatus* decolorized a polymeric dye, Poly-B411 but only when the fungus was previously cultured in lignin containing media (Pagga and Brown, 1986). A 73 kDa peroxidase from *P. ostreatus* has been shown to be involved in Remazol Brilliant Blue decolorization (Shin et al., 1997). Congo Red is readily decolorized by cultures of *Pleurotus ostreiformis* (Dey et al., 1994), and laccases from *Trametes versicolor* can use Remazol Brilliant Blue as a mediator in the oxidation of model lignin compounds (Bourbonnais et al., 1990). In this work, sixteen strains of ligninolytic fungi were examined for the decolorization of twenty-three industrial dyes and an attempt was made to correlate dye decolorization with enzyme production. Two forms of laccase from *Trametes hispida*, shown to be involved in the decolorization reaction, were purified and their kinetic properties were determined.

MATERIALS AND METHODS

Fungal strains: *Bjerkandera adusta* 4312, 7308, 8258; *Pleurotus ostreatus* 7964, 7972, 7980, 7988, 7989, 7992; *Phanerochaete chrysosporium* 3541, 3642; *Sporotichum pulverulentum* 4521; *Trametes hispida* 8260 and *Trametes versicolor* 8272 were obtained from University of Alberta Mold Herbarium, Edmonton, Canada. *Pleurotus ostreatus* IE8 was obtained from the Ecology Institute, Xalapa, Mexico and *Phanerochaete chrysosporium* ATCC 24725 was from the American Type Culture Collection, Rockville Pike, MD. All fungi were maintained on potato dextrose agar plates (Difco). Industrial dyes were obtained from BASF (Germany) and Orisol dyes were obtained from Colorfran S.A. (Mexico).

Inocula were prepared as reported previously (Rodriguez et al., 1999): Fungal mycelium, excised from agar plates, were inoculated in each 125 mL-flasks containing 50 mL of glucose-malt extract-yeast extract medium. After six days of incubation, in shaken flasks at 28°C, the fungal growth in liquid glucose-malt medium reached 2.02 (±0.20) g/L of dry biomass, and 20 mL of this culture was used to inoculate 50 g of wet whole cereal grain at 30°C. After twenty days of mycelial growth in solid state fermentation, extracellular enzymes were extracted by washing three times with 100 mL of 60 mM sodium phosphate buffer, pH 6.0. The combined extracts were filtered and assayed for enzyme activities.

Enzymatic activities, laccase, lignin peroxidase, aryl alcohol oxidase and manganese peroxidase, were determined as previously reported (Rodriguez et al., 1999). One unit of enzyme activity was defined as the amount of enzyme oxidizing 1 μmol of substrate min^{-1}. Decolorization activity was determined by measuring the decrease of the dye absorbance at their maximum visible absorbance wavelength. Dye concentration in the reaction mixture was adjusted to 1.0 absorbance unit at the maximum wavelength in the visible spectrum.

Table 2. Enzymatic activities of crude extracts from solid fungal cultures grown on oat grains (from Rodriguez et al., 1999).

Strains	Volumetric activity (U L⁻¹)[a]			Decolorization activity (ΔA min⁻¹ L⁻¹)[b]				
	Laccase	MnPO[c]	AAO[c]	Reactive blue 158	Acid blue 185	Acid black 194	Orisol blue	Orisol turquoise
B. adusta (4312)[d]	nd[e]	nd	54 - 160	nd	nd	nd	nd	nd
B. adusta (7308)	5 - 7	66 - 70	nd	980 - 1800	520 - 667	nd	nd	133 - 180
B. adusta (8258)	6 - 18	126 - 226	nd	nd	nd	nd	nd	nd
P.ostreatus (7964)	83 - 181	59 - 75	nd	1300 - 2220	440 - 580	nd	nd	560 - 1040
P.ostreatus (7972)	151 - 223	21 - 81	nd	1040 - 3320	140 - 460	0 - 160	0 - 180	400 - 480
P.ostreatus (7980)	109 - 421	47 - 61	24 - 97	8080 - 8420	1440	0 - 100	0 - 400	2040 - 3000
P.ostreatus (7988)	235 - 588	77 - 157	nd	4180 - 10160	1580 - 3500	0 - 360	180 - 600	1800 - 6060
P. ostreatus (7989)	287 - 427	97 - 108	nd	3860 - 6660	880 - 1960	0 - 460	160 - 320	1060 - 2580
P. ostreatus (7992)	134 - 215	78 - 253	41 - 42	1580 - 1820	60 - 600	0 - 40	0 - 260	0 - 780
P. ostreatus (IE8)	403 - 1272	49 - 67	nd	1060 - 9500	1160 - 2500	130 - 500	650 - 900	1200 - 2300
P. chrysosporium (3541)	nd	nd	nd	nd	nd	nd	nd	nd
P. chrysosporium (3642)	nd	nd	nd	nd	nd	nd	nd	nd
P. chrysosporium (ATCC)	nd	nd	nd	nd	nd	nd	nd	nd
S. pulverulentem (4521)	nd	nd	nd	nd	nd	nd	nd	nd
T. hispida (8260)	1184 - 1766	78 - 99	nd	14020 - 21700	2640 - 12840	420 - 740	240 - 780	2840 - 3720
T. versicolor (8272)	86 - 1042	39 - 96	nd	3480 - 14540	400 - 1220	0 - 840	360 - 840	480 - 900

[a] Activity range from three independent replicates. Volumetric activity found in 300 mL extract from fermentation after 20 days growth.
[b] Decolorization activity is estimated as the decrease in absorbance at the maximum visible wavelength for each dye.
[c] MnPO, Manganese peroxidase. AAO, Aryl alcohol oxidase.
[d] Strain number from the University of Alberta Mold Herbarium.
[e] nd. No detected.

Laccase was purified as descrived previously (Rodriguez et al., 1999) from twenty-day-old cultures of solid-state fermentation of *Trametes hispida* 8260 on oats. Laccase I and II showed specific activities of 168 U/mg and 170 U/mg, respectively. Protein concentration was determined by the Bio-Rad protein assay and gel electrophoresis was performed on 10% polyacrylamide gels (SDS-PAGE).

RESULTS AND DISCUSSION

Ligninocellulosic materials were able to induce ligninolytic enzyme production in many fungi (Pickard et al., 1999; Rodirguez et al., 1999). With the aim of finding high decolorization activity, 16 white-rot fungi strains were grown in solid state cultures (Table 1). Oat grains were used as substrate, and lignin peroxidase, manganese peroxidase, laccase and veratryl alcohol oxidase activities were determined in the crude cell-free extracts from all 16 fungal strains. *Trametes hispida* showed the highest laccase activity production, which is consistent with the highest dye decolorization activity. No lignin peroxidase activity could be detected in any of the fungi strains cultured under our growth conditions. All strains of *P. ostreatus* were active to various levels in decolorizing the five dyes tested, but *T. hispida* showed the highest volumetric activity. *Bjerkandera adusta* strains showed high manganese peroxidase but low laccase and decolorizing activity, and *Phanerochaete chrysosporium*, well known as a producer of ligninolytic enzymes under low-nitrogen growth, produced none of the enzymes nor decolorized the dyes under these growth conditions. Lignin peroxidase has been involved in the dye decolorization, mainly in *P. chrysosporium* cultures (Bumpus et al., 1985; Platt et al., 1985; Cripps et al., 1990; Ollikka et al., 1993; Young and Yu, 1997). However, none of the *P. chrysisporium* strains tested under our conditions produced detectable amounts of lignin peroxidase (Table 1), and their extracellular extracts were unable to decolorize indusrial dyes *in vitro*. Manganese peroxidase (Paszczynski and Crawford, 1991; Archibald, 1992) and laccase also have been reported to decolorize some synthetic dyes. From the strains we tested it seems that laccase is the main enzyme involved in dye decolorization. This activity is clearly correlated with the decolorization capacity and the purified preparations are able to perform this decolorization reaction *in vitro*.

Laccases isoenzymes from *T. hispida*, which showed the highest laccase and decolorization activites, were purified and their kinetics constants were determined with ABTS and Reactive blue 158 (Remazol brilliant blue, CI 61200) as substrates (Table 2). Specific activities of the two purified enzymes were 168 U mg^{-1} for laccase I and 170 U mg^{-1} for laccase II. The kinetic constants, determined by double reciprocal plots, showed that the k_{cat} for ABTS oxidation is 10 times higher in laccase II than these found with laccase I. However, the K_M value of laccase II is also higher than of laccase I, making no significant differences in the catalytic efficiency values (k_{cat} / K_M) for ABTS. On the other hand, kinetic constants for the Reactive blue 158 as substrate were similar for both laccases.

Table 2. Kinetic constants of laccase isoenzymes from *Trametes hispida* for ABTS and Rective blue 158 oxidations.

Enzyme	ABTS			Reactive blue		
	k_{cat} (s^{-1})	K_M (μM)	k_{cat}/K_M (s^{-1} mM^{-1})	k_{cat} (s^{-1})	K_M (μM)	k_{cat}/K_M (s^{-1} mM^{-1})
Laccase I	16	16	986	46	3500	13
Laccase II	178	235	759	25	3900	7

Molecular weights determined by SDS-PAGE electrophoresis showed that both laccase I and laccase II are 68 kDa proteins, and both enzymes were stable showing no decrease in activity during 30 days at room temperature under aseptic conditions. These molecular weight are in the same range reported for laccases I and II found in *Pleurotus ostreatus*, 64 kDa (Youn et al., 1995), *Trametes versicolor*, 67 kDa (Bourbonnais and Paice, 1990), *and Pleurotus eryngii*, 65 and 61 kDa respectively (Muñoz et al., 1997). While laccases from different sources have many similar properties, there are also catalytic differences. Although data are presented only for five dyes, *T. hispida* laccase was able to decolorize *in vitro* eleven of the twenty-three industrial, while *P. ostreatus* laccase was only able to oxidize *in vitro* five of same dyes. These results also show that there are other enzymatic systems involved in dye decolorization in *in vivo* cultures, such as cytochromes P450 or peroxidases (Shin et al., 1997).

In conclusion, several industrial dyes were decolorized biocatalytically by extracellular enzymes from different strains of white-rot fungi grown on oats in solid state fermentation. This decolorization capacity was correlated with the laccase activity levels. *Trametes hispida* showed the highest volumetric decolorization activity and purified laccases from *T. hispida* were able to decolorize several synthetic dyes *in vitro*. This enzymatic system appears to be a good candidate for immobilization and use as a bioreactor for effluent treatment from the dye and printing industries.

ACKNOWLEDGMENTS

This work was funded by the Mexican Oil Institute (Grant FIES-108-VI) and by the National Council for Science and Technology of Mexico (Grant 25376-A).

REFERENCES

Archibald F.S. 1992. "A new assay for lignin-type peroxidase employing the dye Azure B." *Appl. Environ. Microbiol.* 58: 3110-3116.

Bourbonnais R. and M.G. Paice 1990. "Oxidation of non-phenolic substrates." *FEBS Lett.* 267: 99-102.

Brown D.H., H.R Hitz. and L. Schafer 1981. "The assesment of the possible inhibitory effect of dyestuffs on aerobic wastewater bacteria. Experience with a screening test." *Chemosphere* 10: 245-261.

Bumpus J.A., M. Tien, D. Wright and S.D. Aust 1985. "Oxidation of persistent environmental pollutants by a white rot fungus." *Science* 228: 1434-1435.

Cripps C., J.A. Bumpus and S.D.Aust 1990. "Biodegradation of Azo and heterocyclic dyes by *Phanerochaete chrysosporium*." *Appl. Environ. Microbiol.* 56: 1114-1118.

Dey S., T.K. Maiti and B.C. Bhattacharyya 1994. "Production of some extracellular enzymes by a lignin peroxidase-producing brown rot fungus, *Pleurotus ostreiformis*, and its comparative abilities for lignin degradation and dye decolorization." *Appl. Environ. Microbiol.* 60: 4216-4218.

Glenn JK and M.H. Gold 1983. "Decolorization of several polymeric dyes by the lignin-degrading basidiomycete *Phanerochaete chrysosporium*." *Appl. Environ. Microbiol.* 45: 1741-1747.

Kirby N., G. Mc Mullan and R. Marchant 1995. "Decolourisation of an artificial textile effluent by *Phanerochaete chrysosporium*." *Biotechnol. Lett.* 17: 761-764.

Muñoz G., F. Gillén, A.T. Martínez and M.J. Marínez 1997. "Laccase isoenzymes of *Pleurotus eryngii* : characterization, catalytic properties, and participation in activation of molecular oxygen and Mn^{+2} oxidation." *Appl. Environ. Microbiol.* 63: 2166-2174.

Ollikka P., K. Alhonmaki, V-M. Leppanen, T. Glumoff, T. Raijola, Y. Suominen 1993. Decolorization of azo, triphenyl methane, heterocyclic, and polymeric dyes by lignin peroxidase isoenzymes from *Phanerochaete chrysosporium*. Appl. Environ. Microbiol. 59: 4010-4016.

Pickard M.A., H. Vandertol, R. Roman and R. Vazquez-Duhalt 1999. "High production of ligninolytic enzymes from white rot fungi in cereal bran liquid medium." *Can. J. Microbiol.* (in press)

Platt M.W., Y. Hadar and H. Chet 1985. "The decolorization of the polymeric dye Poly-blue (polyvinalamine sulfonate-anthraquinone) by lignin degrading fungi." *Appl. Microbiol. Biotechnol.* 21: 394-396.

Pagga U. and D.H. Brown 1986. "The degradation of dyestuffs part II. Behavior of the dyestuffs in aerobic biodegradation test." *Chemosphere* 15: 479-491.

Pasti-Grigsby M.B., A. Paszczynski, S. Goszcynski, D.L. Crawford and R.L. Crawford 1992. "Influence of aromatic substitution patterns on azo dye degradability by *Streptomyces spp.* and *Phanerochaete chrysosporium.*" Appl. Environ. Microbiol. 58: 3605-3613.

Paszczynski A. and R.L. Crawford 1991. "Degradation of azo compounds by ligninase from *Phanerochete chrysosporium*: involvement of veratryl alcohol." *Biochem. Biophys. Res. Comm.* 178: 1056-1063.

Rodriguez E., M.A. Pickard. and R. Vazquez-Duhalt 1999. "Industrial dye decolorization by laccases from ligninolytic fungi." *Current Microbiol.* 38: 27-32.

Shin K.S., I.K. Oh and C.J. Kim 1997. "Production and purification of Remazol Brilliant Blue R decolorization peroxidase from the culture filtrate of *Pleurotus ostreatus.*" *Appl. Environ. Microbiol.* 63: 1744-1748.

Youn H.D., K.J. Kim, Y.H. Han, I.B. Jeong, G. Jeong, S.O. Kang, Y.C.Hah 1995. "Single electron transfer by an extracellular laccase from the white-rot fungus *Pleurotus ostreatus.*" *Microbiology* 141: 393-398.

Young L and J. Yu 1997. "Ligninase-catalyzed decolorization of synthetic dyes". *Water Res.* 31: 1187-1193.

AUTHOR INDEX

This index contains names, affiliations, and volume/page citations for all authors who contributed to the eight-volume proceedings of the Fifth International In Situ and On-Site Bioremediation Symposium (San Diego, California, April 19–22, 1999). Ordering information is provided on the back cover of this book. The citations reference the eight volumes as follows:

5(1): Alleman, B.C., and A. Leeson (Eds.), *Natural Attenuation of Chlorinated Solvents, Petroleum Hydrocarbons, and Other Organic Compounds*. Battelle Press, Columbus, OH, 1999. 402 pp.

5(2): Leeson, A., and B.C. Alleman (Eds.), *Engineered Approaches for In Situ Bioremediation of Chlorinated Solvent Contamination*. Battelle Press, Columbus, OH, 1999. 336 pp.

5(3): Alleman, B.C., and A. Leeson (Eds.), *In Situ Bioremediation of Petroleum Hydrocarbon and Other Organic Compounds*. Battelle Press, Columbus, OH, 1999. 588 pp.

5(4): Leeson, A., and B.C. Alleman (Eds.), *Bioremediation of Metals and Inorganic Compounds*. Battelle Press, Columbus, OH, 1999. 190 pp.

5(5): Alleman, B.C., and A. Leeson (Eds.), *Bioreactor and Ex Situ Biological Treatment Technologies*. Battelle Press, Columbus, OH, 1999. 256 pp.

5(6): Leeson, A., and B.C. Alleman (Eds.), *Phytoremediation and Innovative Strategies for Specialized Remedial Applications*. Battelle Press, Columbus, OH, 1999. 340 pp.

5(7): Alleman, B.C., and A. Leeson (Eds.), *Bioremediation of Nitroaromatic and Haloaromatic Compounds*. Battelle Press, Columbus, OH, 1999. 302 pp.

5(8): Leeson, A., and B.C. Alleman (Eds.), *Bioremediation Technologies for Polycyclic Aromatic Hydrocarbon Compounds*. Battelle Press, Columbus, OH, 1999. 358 pp.

KEYWORD INDEX

This index contains keyword terms assigned to the articles in the eight-volume proceedings of the Fifth International In Situ and On-Site Bioremediation Symposium (San Diego, California, April 19-22, 1999). Ordering information is provided on the back cover of this book.

In assigning the terms that appear in this index, no attempt was made to reference all subjects addressed. Instead, terms were assigned to each article to reflect the primary topics covered by that article. Authors' suggestions were taken into consideration and expanded or revised as necessary. The citations reference the eight volumes as follows:

5(1): Alleman, B.C., and A. Leeson (Eds.), *Natural Attenuation of Chlorinated Solvents, Petroleum Hydrocarbons, and Other Organic Compounds.* Battelle Press, Columbus, OH, 1999. 402 pp.

5(2): Leeson, A., and B.C. Alleman (Eds.), *Engineered Approaches for In Situ Bioremediation of Chlorinated Solvent Contamination.* Battelle Press, Columbus, OH, 1999. 336 pp.

5(3): Alleman, B.C., and A. Leeson (Eds.), *In Situ Bioremediation of Petroleum Hydrocarbon and Other Organic Compounds.* Battelle Press, Columbus, OH, 1999. 588 pp.

5(4): Leeson, A., and B.C. Alleman (Eds.), *Bioremediation of Metals and Inorganic Compounds.* Battelle Press, Columbus, OH, 1999. 190 pp.

5(5): Alleman, B.C., and A. Leeson (Eds.), *Bioreactor and Ex Situ Biological Treatment Technologies.* Battelle Press, Columbus, OH, 1999. 256 pp.

5(6): Leeson, A., and B.C. Alleman (Eds.), *Phytoremediation and Innovative Strategies for Specialized Remedial Applications.* Battelle Press, Columbus, OH, 1999. 340 pp.

5(7): Alleman, B.C., and A. Leeson (Eds.), *Bioremediation of Nitroaromatic and Haloaromatic Compounds.* Battelle Press, Columbus, OH, 1999. 302 pp.

5(8): Leeson, A., and B.C. Alleman (Eds.), *Bioremediation Technologies for Polycyclic Aromatic Hydrocarbon Compounds.* Battelle Press, Columbus, OH, 1999. 358 pp.